TEAM VISIO

THE VISUAL ATLAS OF MYSTERIOUS CREATURES

Monsters, Cryptids, Sightings, and Unsolved Mysteries

intra

Visio Series

Edizioni Intra®
www.intrapublishing.com
info@intra.pro

ISBN 979-12-5991-841-3

Contents

CONTENTS 3

METHODOLOGICAL PREMISE. 5

THE SCIENCE OF THE IMPOSSIBLE: BETWEEN CRYPTOZOOLOGY, PAREIDOLIA, AND REAL DISCOVERIES 5

SECTION I. HOMINIDS AND FRONTIER ANTHROPOLOGY 9

- Bigfoot (Sasquatch) 10
- Yeti 15
- Yeren 22
- Dahu 27

SECTION II. AQUATIC CRYPTIDS AND UNEXPLORED DEPTHS 31

- Loch Ness (Nessie) 32
- Ogopogo 39
- Lariosaurus 43
- Kraken & Globster 48
- Mokele-Mbembe 53

SECTION III. AERIAL AND ATMOSPHERIC PHENOMENA 57

- Mothman 58
- Thunderbird 63
- Jersey Devil 67

SECTION IV. SPECULATIVE BIOLOGY AND HYBRIDS 73

- Chupacabra 74
- Tatzelwurm 78
- Tsuchinoko 82
- Flatwoods Monster 86
- Thylacine (Tasmanian Tiger) 91

APPENDIX. THE GREAT ABSENTEES 95

LEGAL NOTICE 99

Methodological Premise.
The Science of the Impossible: Between Cryptozoology, Pareidolia, and Real Discoveries

The zoological map of our planet now appears as a dense and complete grid, where every creature seems to have found its proper taxonomic place and every corner of the forest appears to have been scanned by satellites. Yet, at the margins of this reassuring map, a stubborn shadow zone survives—a borderland where biology fades into folklore. It is here that cryptozoology operates, a term formalized in the late 1950s (the first documented occurrence dates to 1959) by the Belgian zoologist Bernard Heuvelmans to define the study of "hidden animals." Unlike mythology, which deals with symbols and archetypes, cryptozoology advances a purely biological hypothesis: that large animal species not yet cataloged by official science may still exist, continuing to evade direct human observation, leaving behind only fleeting traces, anecdotes, and controversial evidence.

The arrogance of academic science has often declared discoveries impossible long before their time. The celebrated naturalist Georges Cuvier, at the beginning of the nineteenth century, firmly stated that there was no hope of discovering new large animal species. History quickly proved him wrong. Throughout the nineteenth and twentieth centuries, creatures once considered legendary or extinct emerged from the mists of myth to enter university textbooks. The most emblematic case is that of the okapi, a relative of the giraffe that inhabits the dense forests of the Congo: for decades, European explorers dismissed it as a kind of "African unicorn" based on local accounts, until in 1901 its skin reached London, forcing zoology to revise its lists.

Yet the true totem for those who investigate the unknown is the coelacanth (Latimeria chalumnae). This prehistoric fish was known to paleontologists only through fossils and was believed to have been extinct for more than sixty-five million years, having disappeared alongside the dinosaurs at the end of the Cretaceous period. Its rediscovery in 1938—alive and well in the nets of a fishing vessel off the coast of South Africa—introduced the concept of the "Lazarus taxon": organisms that return from the realm of the dead. If a six-foot fish managed to remain hidden for millions of years in the ocean depths, it is legitimate to ask whether the vast forests of North America or the dark lakes of Scotland might conceal equally astonishing surprises.

The investigation of modern cryptids, however, cannot rely solely on biological hope; it must also confront the fallibility of human perception. When we analyze sightings of creatures such as Bigfoot or the Mothman, we enter the fields of eyewitness psychology and neuroscience. The human brain is a machine designed for pattern recognition: evolution has programmed us to identify faces and figures even where none exist, a phenomenon known as pareidolia. Under conditions of poor visibility, emotional stress, or fear, a tree trunk swaying in the wind becomes a humanoid figure, and a rogue wave transforms into the back of a sea monster.

Added to this is the cultural filter: we see what we expect to see. In the Middle Ages, travelers glimpsed dragons and basilisks; in the atomic and space age, entities have taken on more mechanical, alien, or mutant forms. This volume does not intend to ridicule such testimonies, but to analyze them with the cold detachment of forensic investigation. Each chapter that follows will treat a creature not as a movie monster, but as an "open case." We will examine the physical evidence, assess the ecological plausibility of a breeding population, and attempt to separate the background noise of legend from the faint but persistent signal of a possible biological reality yet to be discovered. For, as the history of science teaches us, absence of evidence is never definitively evidence of absence.

READING NOTES

To enrich the reading experience and make it more immersive, this book includes QR codes that provide immediate access to external digital content. A QR code is a two-dimensional barcode that, when scanned with the camera of a smartphone, tablet, or computer, automatically opens a link.

In particular, for each entry you will find several main QR codes:

- Google Maps
- Google Earth
- YouTube
- JustWatch
- Other Web Sites

Below each QR code, only the platform's logo is displayed, ensuring a simple, quick, and visually clean user experience.

To use this feature correctly:
1. Open your device's camera and frame the QR code.
2. If your device supports automatic QR recognition, a clickable link will appear.
3. Alternatively, you can use a free QR code reader app, easily available in digital app stores.

SECTION I.
Hominids and Frontier Anthropology

Bigfoot (Sasquatch)

The modern history of what is widely considered the absolute icon of cryptozoology officially begins—rather prosaically—on August 27, 1958. Not in a university laboratory, but at a logging site in Humboldt County, Northern California. On that day, a bulldozer operator named Jerry Crew walked into the newsroom of the local newspaper, the Humboldt Times, carrying a plaster cast of a monstrous footprint: sixteen inches long and five inches wide, discovered in the mud near Bluff Creek. Journalist Andrew Genzoli, sensing a front-page story, coined the term "Bigfoot" in his article. Postwar America, hungry for mysteries unrelated to the Cold War, immediately embraced the idea of a giant, hairy being hidden among the redwoods. Yet reducing the phenomenon to a 1950s media stunt would be a historical and anthropological mistake.

Long before Jerry Crew poured plaster into the mud, the Indigenous peoples of the Pacific Northwest lived with the certainty that they were not the only bipeds walking those forests. The term Sasquatch, now used as a more respectful and quasi-scientific synonym for Bigfoot, is an anglicization of the word Sásq'ets from the Halkomelem language spoken by tribes of the Fraser Valley in British Columbia. As early as the 1920s, Indian agent J.W. Burns collected dozens of oral testimonies describing the "wild men of the woods": not immaterial spirits, but flesh-and-blood beings, foul-smelling, whistling to communicate, and stealing dried salmon. The overlap between ancestral legends and the sightings reported by European settlers suggests a narrative continuity that challenges the idea of simple folkloric invention. Instead, it outlines the profile of a large omnivorous primate, primarily nocturnal, that may have found refuge in the vast, nearly unexplored wilderness of the Cascade Mountains.

The turning point—the event that transformed Sasquatch from a local myth into a global "case"—occurred nine years after Jerry Crew's discovery, in the very same place: Bluff Creek. On the afternoon of October 20, 1967, two former rodeo cowboys, Roger Patterson and Robert Gimlin, were riding along the creek bed when they encountered a massive figure crouched near the water. Their horses panicked. Patterson fell, grabbed his 16mm Cine-Kodak K-100 camera, and ran toward the creature, managing to film for just under a minute a bipedal being covered in dark hair, walking away with a decisive yet fluid stride toward the tree line. That single minute of grainy footage remains, nearly sixty years later, the most controversial and analyzed piece of visual evidence in the history of frontier zoology.

Academic science almost immediately dismissed the film as a hoax: a man in a costume. Yet analyses conducted in subsequent decades using increasingly sophisticated digital technologies have revealed anatomical details that make the costume hypothesis highly problematic given the state of special effects in 1967. Experts in biomechanics and physical anthropology, such as Professor Grover Krantz of Washington State University and later Dr. Jeff Meldrum of Idaho State University, have observed that the subject in the film—nicknamed "Patty"—walks with what is termed a "compliant gait." The knees remain flexed to absorb the estimated weight—approximately 650 pounds—in a manner distinct from modern humans, who tend to lock the knee with each step. Image stabilization further reveals muscle masses contracting and relaxing beneath the hair, particularly in the thigh and back, as well as a visible oscillation of breast tissue, suggesting the specimen may be female.

The matter remains an unresolved scientific paradox. On one hand, the absence of a body, bones, or fossil remains in the North American record makes taxonomic classification impossible—although a possible relationship to Gigantopithecus blacki, an enormous Asian primate believed by paleoanthropologists to have gone extinct approximately 300,000 years ago, is frequently proposed. On the other hand, the sheer volume of circumstantial evidence—thousands of footprints displaying dermal ridges, audio recordings known as the "Sierra Sounds" exhibiting vocal ranges beyond human capacity, and forensic analyses of the Patterson-Gimlin film—prevents the file from being definitively closed. Bigfoot remains a biological shadow walking the boundary between evolutionary reality and the deeply human desire for a world that is still wild and unexplored.

BFRO (Bigfoot Field Researchers Organization)

Frame 352 from Patterson Gimlin Bigfoot video (1967)

12N42, Happy Camp, CA 96039, USA	41°26'25"N 123°42'06"W	Patterson-Gimlin film (1967)
Bigfoot Videos and Documentaries	Bigfoot Sounds and Recordings	Bigfoot Film

Bigfoot (Sasquatch) sighting in a conifer forest
Artificially generated image for illustrative and narrative purposes

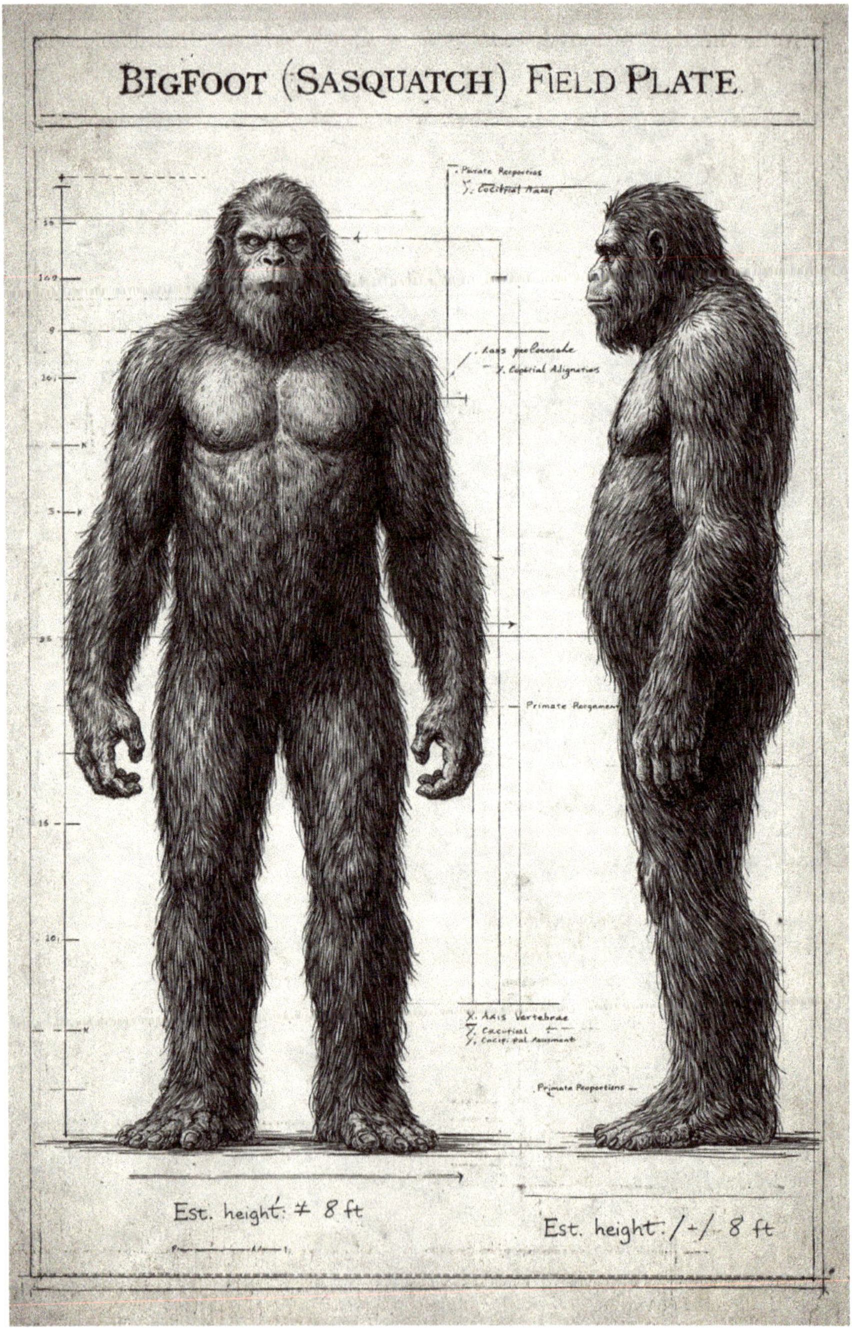

Artificially generated image for illustrative and narrative purposes

Yeti

Among all the creatures that populate the atlas of the unknown, the Yeti holds a position of absolute primacy in the collective imagination, rivaling Bigfoot for the title of the world's most famous cryptid. Yet its Western fame rests on a linguistic and visual misunderstanding that distorted, for decades, the reality of a centuries-old Himalayan myth. The label "Abominable Snowman," evoking the image of a bloodthirsty monster with white fur like a bipedal polar bear, is in fact the product of sensationalist journalism in 1921. That year, Lieutenant Colonel Charles Howard-Bury, returning from a reconnaissance expedition on Everest, reported finding footprints that his Sherpa porters attributed to the Metoh-Kangmi. The term, roughly meaning "man-bear of the snows," was mistranslated and dramatized by journalist Henry Newman of The Calcutta Statesman as "abominable," transforming a figure of local folklore into a global Gothic nightmare.

The true nature of the Yeti — or Yeh-teh, "that thing over there" in local languages — differs greatly from its cinematic caricature. The peoples of Tibet and Nepal do not describe an ice monster, but a dark creature with brown or reddish fur that inhabits not the sterile summits of perpetual snow, but the forested belt of rhododendrons and conifers, venturing onto glaciers only occasionally to cross high passes. It is a liminal being, suspended between the animal and human worlds, often depicted in local religious art as a guardian of the mountains. The physical evidence that elevated the Yeti from superstition to zoological problem emerged on November 8, 1951, on the Menlung Glacier at over 20,000 feet of elevation. British explorer Eric Shipton photographed a fresh footprint in the snow: more than twelve inches long, broad and squat, it displayed an enormous big toe clearly separated from the others — a morphology disturbingly reminiscent of an anthropoid primate and unlike any known high-altitude fauna.

That single image, sharp and unsettling, ignited the "Yeti-mania" of the 1950s, culminating in several scientific expeditions, including one led in 1960 by Sir Edmund Hillary. The goal was to obtain definitive biological proof. Buddhist monasteries in the Khumbu region, such as Khumjung, had for generations preserved supposed relics of the creature: conical scalps of dark hair and mummified hands. Analyses conducted at the time and in subsequent decades deflated much of the enthusiasm: the scalps proved to be pieces of serow hide (a local goat-antelope) shaped into ceremonial objects, and the hands were assemblages of human and animal bones. It seemed the end of the myth. And yet eyewitness accounts from respected climbers — including Reinhold Messner — continued to describe encounters with a large unidentified biped.

The decisive breakthrough, which may have resolved the enigma without stripping it of its fascination, came from twenty-first-century molecular genetics. Several genetic studies, including pioneering work by Bryan Sykes (University of Oxford) and more recent analyses by Professor Charlotte Lindqvist (University at Buffalo), examined the mitochondrial DNA of dozens of samples. The results pointed almost unequivocally toward two local bear subspecies: the Himalayan black bear (Ursus thibetanus laniger) and, especially, the Himalayan brown bear (Ursus arctos isabellinus). The latter is an extremely rare and elusive animal, genetically distinct from other brown bears due to long isolation in mountain ranges for hundreds of thousands of years. It is plantigrade and capable of standing upright on its hind legs for extended periods, appearing from a distance like a dark, shaggy giant walking like a man. In this sense, science has not erased the Yeti but given it a Latin name: the creature exists, it is real, and it is a living fossil of ursid evolution, transformed by human culture into the solitary guardian of the peaks.

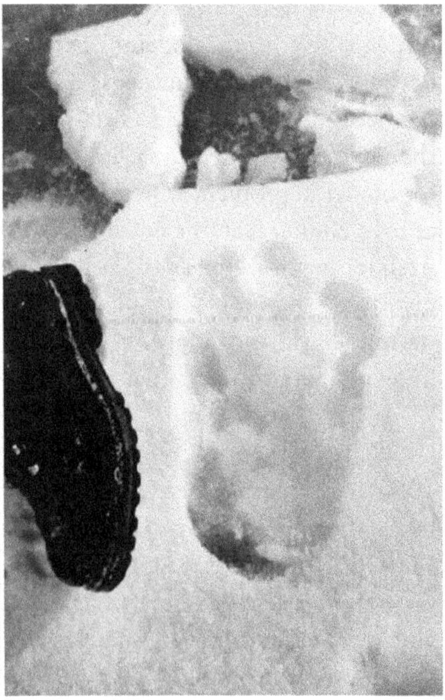

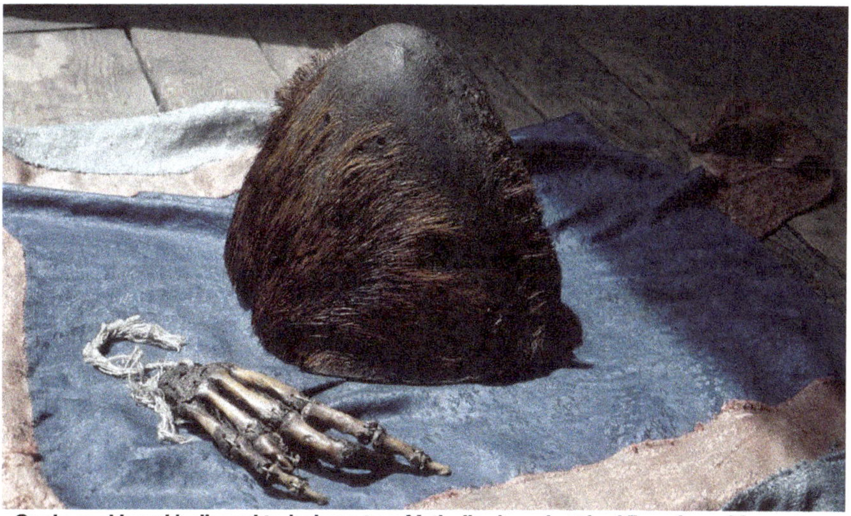

Scalp and hand believed to belong to a Yeti, displayed at the Khumjung Monastery (Nepal) – Source: Traveller

Khumjung Monastery RPG8+285, Khumjung 56000, Nepal	27°49'30"N 86°42'58"E	Khumjung Monastery Yeti scalp
Yeti Videos and Documentaries	Eric Shipton's Everest reconnaissance (1951)	Yeti – Film

Photograph from Eric Shipton's 1951 reconnaissance – Source: Hornseys

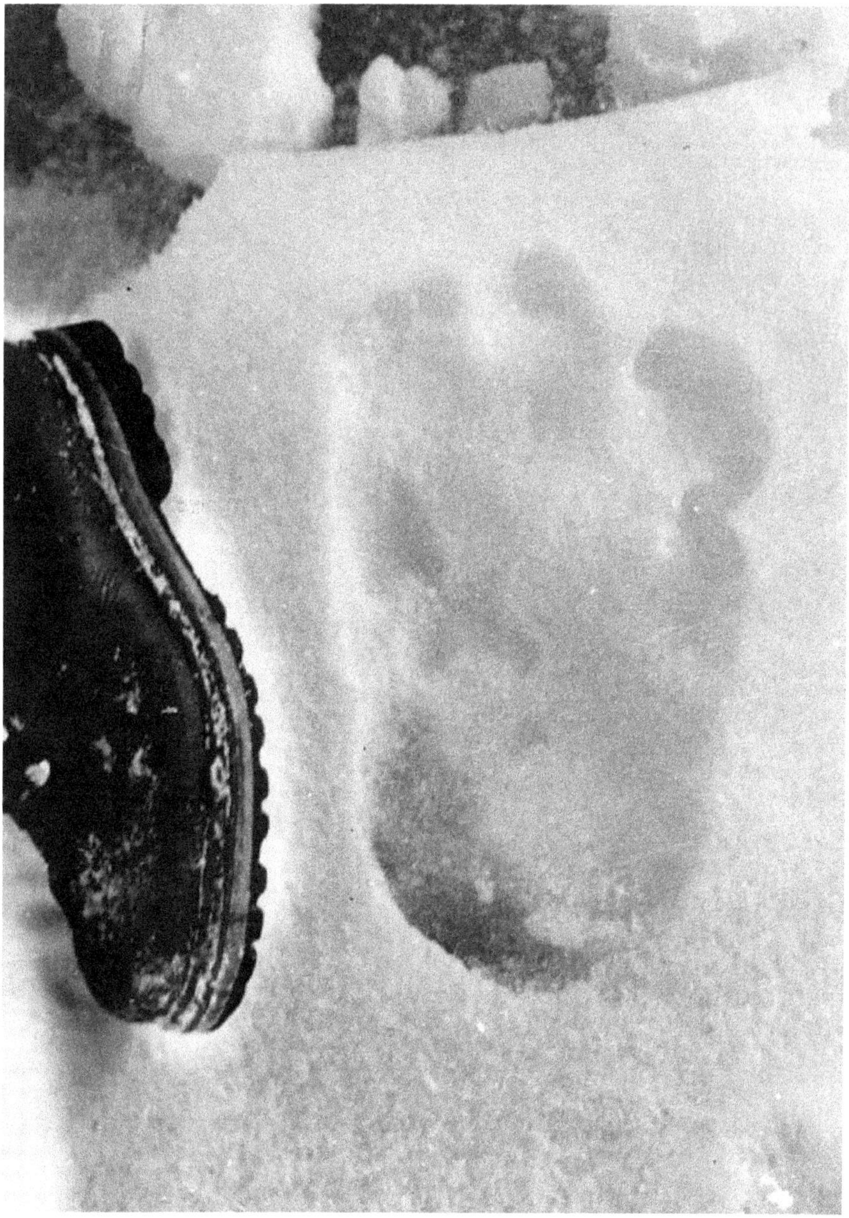

The footprint photographed by Eric Shipton in 1951. To provide a sense of scale, Shipton placed his ice axe beside it.
Source: Hornseys

Artificially generated image for illustrative and narrative purposes

Yeren

While in the West the search for Bigfoot has historically been left to amateur hunters and eccentric enthusiasts—often ridiculed by academia—in China the question of the Wild Man has, for decades, enjoyed a surprising degree of institutional legitimacy. The Yeren, or "Wild Man" (Ye Ren), is not merely a campfire ghost story, but a paleoanthropological enigma rooted in the vast expanse of the Shennongjia Nature Reserve in central Hubei Province. This region, a labyrinth of limestone peaks and primordial forests of conifers and bamboo often shrouded in mist, is considered one of the planet's major biodiversity refuges—a place where the existence of an unclassified species appears, in the eyes of many Chinese scientists, a statistically plausible possibility.

The historical continuity of sightings sets the Yeren apart from many modern cryptids. As early as the third century BCE, the poet Qu Yuan, himself a native of Hubei, described in his verses "mountain spirits" with human-like features. Yet it was in the second half of the twentieth century that the myth became a state objective. Between 1974 and 1981, the Chinese Academy of Sciences (CAS) organized and funded large-scale official expeditions, deploying hundreds of researchers, biologists, and soldiers from the People's Liberation Army to comb the jungles of Shennongjia. The composite profile that emerged from local testimony and military reports is consistent: a biped approximately six and a half feet tall, covered in thick fur ranging from dark brown to bright reddish tones, with an almost human face—prognathous and tailless.

Although the expeditions never captured a living specimen nor recovered a body, they produced a substantial body of circumstantial evidence that keeps the debate open. Casts of footprints up to sixteen inches long were collected, as were woven ground nests resembling those of African great apes and, most notably, thousands of individual hair samples. Microscopic and molecular analyses conducted on these samples often yielded frustrating or ambiguous results: many belonged to known species such as the golden snub-nosed monkey or the Asiatic black bear, yet a percentage displayed medullary structures and pigmentation ratios not matching any known regional animal, while approximating characteristics found in higher primates.

The most compelling scientific theory, advanced by several Chinese and Western paleoanthropologists, directly links the Yeren to the local fossil record. Southern China was, for millions of years, the habitat of Gigantopithecus blacki, the largest ape ever to walk the Earth—a giant that may have reached heights of ten feet. Fossils of Gigantopithecus, consisting primarily of enormous mandibles and teeth, confirm its presence in the region until approximately 300,000 years ago. Is it possible that a relict population of these massive primates, reduced in size to adapt to dwindling resources and competition from Homo sapiens, survived in the isolated pockets of Hubei's mountains? From this perspective, the Yeren would cease to be a monster and instead become a "living fossil," the last heir to a dynasty of giants that once shared Pleistocene forests with our ancestors.

Today, as Shennongjia has become a UNESCO World Heritage Site and a tourist destination, research is less militarized but continues quietly—driven by the awareness that in those vertical valleys, extinction may not have been final.

G88J+WVF,
Shennongjia, Hubei,
Cina, 442421

31°31'02"N 110°19'55"E

Yeren
Videos and
Documentaries

Is China's 'Bigfoot' real?
- Finding the yeren

Yeren – Film

Artificially generated image for illustrative and narrative purposes

Artificially generated image for illustrative and narrative purposes

Artificially generated image for illustrative and narrative purposes

Dahu

Among the rocky folds of the Alps, where the steepness of the slopes challenges both gravity and mammalian endurance, survives a creature that represents perhaps the most refined paradox of European cryptozoology: the Dahu. Unlike the great American or Asian primates, whose existence is debated on the basis of physical evidence, the Dahu occupies an ecological niche suspended between fantastic zoology and cultural anthropology. Widespread throughout the Alpine arc—from the Aosta Valley to Trentino, through the Swiss cantons and the valleys of French Savoy—this animal is consistently described as an ungulate resembling an ibex or chamois, distinguished by a dramatic morphological asymmetry: the legs on one side are significantly shorter than those on the other.

This anatomical peculiarity, far from being a random deformity, is described by folkloric naturalists as an extreme evolutionary adaptation to a vertical environment. The Dahu rupicapra asymmetrica (according to one of the pseudo-scientific classifications proposed in the nineteenth century) would be the perfect animal for mountain slopes, yet condemned by its own specialization to a one-directional existence. The creature can move comfortably only by keeping its shorter legs uphill and its longer legs downhill; attempting to turn around or change direction on level ground would mean losing balance and falling. This characteristic has led to the theorization of two distinct subspecies, destined never to meet head-on: the levogiro Dahu, with shorter left legs, circling the mountain counterclockwise; and the destrogiro Dahu, with shortened right legs, moving clockwise. According to the tales of old hunters, any encounter between the two would end in tragic—or comically impossible—results, making mating a feat of high acrobatic engineering.

Although academic science has never cataloged Dahu bones, dismissing the phenomenon as a playful fable invented by mountaineers to tease naïve city dwellers, an analysis of the myth reveals unexpected depth. The Dahu is not merely a prank; it is the symbolic embodiment of the harshness of Alpine life, a biological metaphor for those forced to adapt to an oblique and hostile world. The famous "Dahu hunt"—a ritual in which a novice is persuaded to wait in the cold with an open sack while beaters pretend to drive the animal toward him—possesses all the characteristics of an initiatory rite of passage. It serves to cement the community, distinguishing those who "belong to the mountain" and understand its unwritten rules from those who do not.

Yet there exists a perspective of speculative biology that makes the Dahu less absurd than it first appears. In nature, extreme morphological adaptation is a reality: consider the ibex (Capra ibex), whose hooves are divided into two movable parts with sharp edges and soft, anti-slip pads—a biological technology that allows it to scale near-vertical cliffs—or the fiddler crab with its disproportionate claw. The Dahu is the dreamlike exaggeration of these principles, a collective thought experiment that has generated a "phantom" so deeply rooted in local culture that it has inspired dedicated museums and symbolic protection statutes. Investigating the Dahu means exploring how mountain dwellers projected their physical experience of steepness onto the animal kingdom, creating a creature that, though nonexistent in flesh, is undeniably alive in the collective memory of the Alps.

Val d'Hérens, 1983 Evolène, Svizzera	46°08'06"N 7°28'10"E	Dahu Videos and Documentaries

Artificially generated image for illustrative and narrative purposes

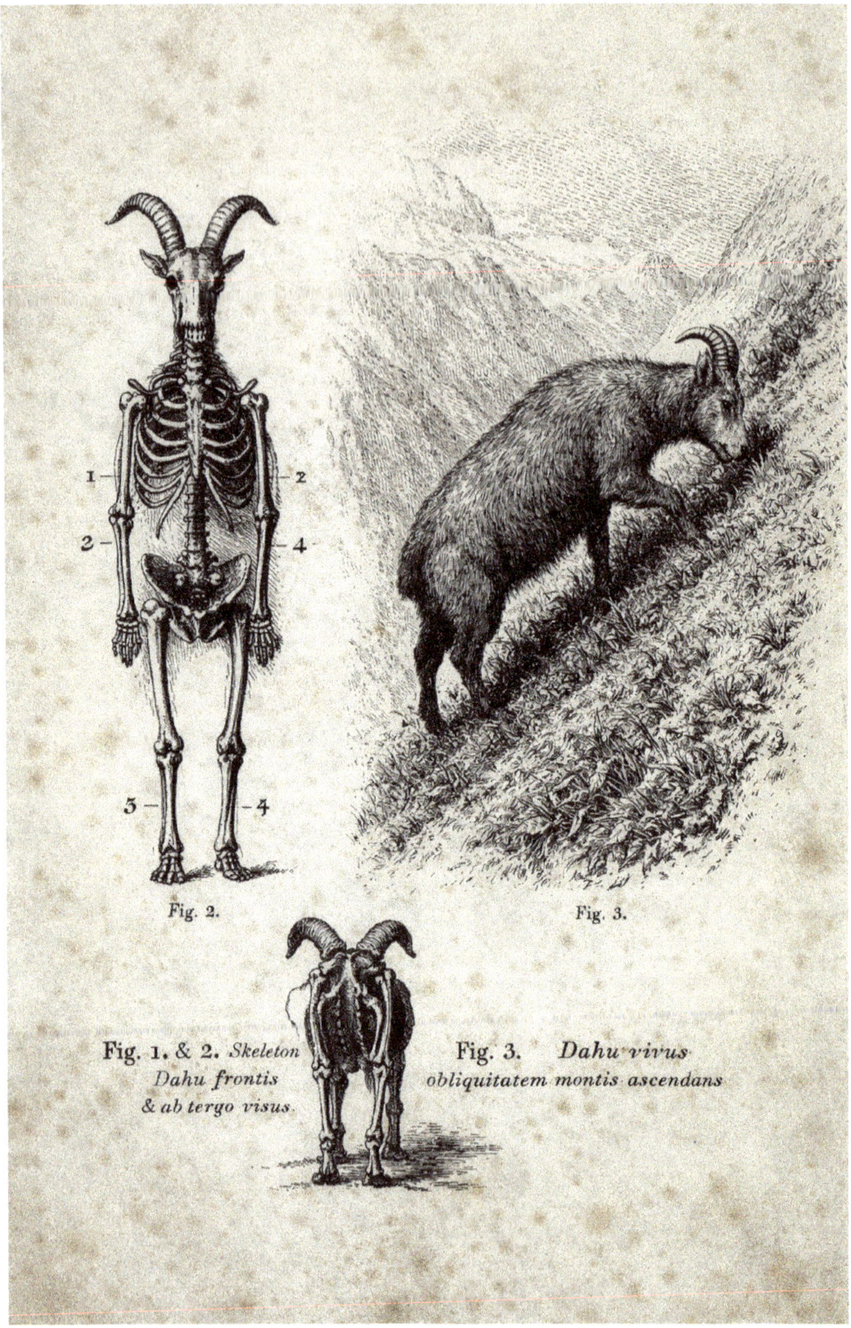

Artificially generated image for illustrative and narrative purposes

SECTION II.
Aquatic Cryptids and Unexplored Depths

Loch Ness (Nessie)

To understand the mystery of Loch Ness, one must first question the land that contains it, even before scanning the water's surface. The lake is not merely a body of water, but a deep and ancient geological scar: the Great Glen Fault. This enormous tectonic fracture cuts diagonally across the Scottish Highlands, creating a rift that hosts a chain of lakes, of which Loch Ness is the largest. Stretching thirty-seven kilometers in length and plunging to depths of approximately 230 meters, it holds more freshwater than all the lakes of England and Wales combined. It is an alien, almost abyssal environment: the waters remain at a constant and frigid temperature of about 41°F (5°C), and underwater visibility is virtually nonexistent. The exceptionally high concentration of suspended peat, washed down from the surrounding mountains, turns the water black as ink just a few feet below the surface. It is within this setting of perpetual darkness—where sunlight barely penetrates—that humanity has projected its fears and hopes of discovery for centuries.

Although the first mention of an "aquatic beast" dates back to 565 A.D. in the Vita Sancti Columbae (in which the Irish saint is said to have driven away a monster in the River Ness, not the lake), the modern phenomenon of "Nessie" has a precise birthdate: 1933. Until then, the loch had remained isolated, accessible only via rough trails. The completion of the A82 road along its northern shore changed everything, offering for the first time a continuous and unobstructed view of the water. In April of that year, the Mackay couple, who managed a local hotel, reported seeing an "enormous beast rolling and plunging" in the lake. The story, amplified by the Inverness Courier, triggered a chain reaction. Within months, the Highlands were flooded with journalists, big-game hunters, and tourists armed with binoculars, transforming a quiet Scottish valley into the global epicenter of cryptozoology.

The image that today defines Nessie—the long, elegant neck rising from the water like that of a plesiosaur that survived the Cretaceous extinction—was cemented in 1934. The so-called "Surgeon's Photograph," allegedly taken by the respected London gynecologist Robert Kenneth Wilson, showed an unmistakable silhouette against the rippling surface of the loch. For sixty years, that photograph stood as the flagship evidence for believers: proof that a prehistoric reptile swam in Scottish waters. Only in 1994 did the truth emerge, following Christian Spurling's deathbed confession: the image was an ingenious and handcrafted hoax. Not a ten-meter monster, but a 12-inch toy submarine topped with a neck sculpted from wood putty, constructed to avenge a newspaper that had ridiculed Spurling's stepfather. The collapse of the "Surgeon's Photograph" dealt a severe blow to the myth's credibility, yet paradoxically did not halt its momentum. The search had already shifted from fallible visual sightings—susceptible to pareidolia—to technological investigation.

In the 1970s, the Academy of Applied Science of Boston, led by Dr. Robert Rines, brought sonar and underwater strobe photography to the loch. In 1972 and 1975, Rines's equipment captured grainy images that appeared to show a giant rhomboid flipper and, in one instance, a grotesque head resembling a gargoyle. Based on these images, naturalist Sir Peter Scott assigned the creature a scientific name, Nessiteras rhombopteryx ("the Ness monster with the diamond-shaped fin"), a move that would have granted the species legal protection under British law. Subsequent analyses, however, suggested that the "head" might have been a submerged tree trunk and the "flipper" a digital processing artifact.

The most systematic and spectacular attempt to "drain" the loch of its secrets occurred in October 1987 with Operation Deepscan. A flotilla of twenty-four boats equipped with state-of-the-art Lowrance echo sounders sailed in tight formation along the entire length of the lake, creating an impenetrable "sonar curtain." The operation cost approximately one million pounds and drew global media attention. The results were ambiguous: sonar registered three large contacts at considerable depth—objects "larger than a shark but smaller than a whale" that appeared to move. Unfortunately, visual identification proved impossible. Skeptics proposed shoals of salmon or seals that had entered from the sea; believers saw in the "Deepscan contacts" definitive proof that something massive inhabited the depths.

In the twenty-first century, the hunt has changed radically. No longer searching for shapes, researchers now seek invisible traces. In 2018, New Zealand geneticist Neil Gemmell of the University of Otago led an ambitious project based on environmental DNA (eDNA). Every aquatic creature leaves behind biological traces: scales, skin cells, feces, urine. By sequencing genetic material extracted from 250 water samples collected at various depths, Gemmell's team created a comprehensive census of life in the loch. The results, published in 2019, were devastating for the "Lost World" hypothesis: no reptile DNA (neither plesiosaur nor otherwise), no shark or giant sturgeon traces. However, the study revealed a surprising abundance of European eel (Anguilla anguilla) DNA.

This finding revived an old theory: Nessie might not be a dinosaur, but a giant eel. Eels are known for their phenotypic plasticity and can reach considerable sizes. Is it biologically possible—though rare—that sterile individuals fail to migrate to the Sargasso Sea to reproduce and die, instead remaining in the loch and continuing to grow indefinitely? An eel as thick as a human thigh and several meters long, swimming near the surface, could easily be mistaken for the neck or hump of a monster.

Today, Loch Ness remains a site of secular pilgrimage. Even without a body to display, the myth generates tens of millions of pounds annually in economic revenue. Whether it is a hypertrophic eel, a standing wave phenomenon (seiche) deceiving the eye, or simply humanity's desire to believe that something vast and wild still eludes us, Nessie continues to swim undisturbed in the dark waters of our collective imagination.

The "Surgeon's Photograph" (1934), now known to be a hoax

Loch Ness – Scotland, United Kingdom

Loch Ness 57°19'35"N 4°25'24"W

Urquhart Castle (observation point) 57°19'27"N 4°26'31"W

Nessie on the Net Live Cam

Operation Deepscan 1987 - Documentaries

Loch Ness Videos and Documentaries

Prof. Neil Gemmell Environmental DNA study of Loch Ness

Loch Ness Centre & Exhibition Drumnadrochit

Loch Ness Film

Artificially generated image for illustrative and narrative purposes

Artificially generated image for illustrative and narrative purposes

Artificially generated image for illustrative and narrative purposes

Ogopogo

If the Loch Ness Monster is the media celebrity of the world of lake cryptids, Ogopogo represents its older, better-documented, and in some respects more unsettling counterpart. In the deep, cold waters of Lake Okanagan, in Canada's British Columbia, a mystery has unfolded for centuries—long before the arrival of European settlers. Unlike the Scottish legend, which exploded as a tourist phenomenon only in the 1930s, the creature of Lake Okanagan is rooted in the spirituality of the First Nations since time immemorial. The Syilx people know the entity as Naitaka (or N'ha-a-itk), a term roughly translated as "the lake demon" or "sacred water spirit." For Indigenous communities, it was not a monster to be hunted but a force of nature to be respected and feared. It was customary to offer tobacco or small animals to the waters before crossing the lake by canoe, especially near Rattlesnake Island, believed to be the creature's lair.

The transition from spiritual entity to tabloid monster occurred in the 1920s, when colonial culture trivialized the myth by renaming it "Ogopogo," borrowing the name from a popular English music-hall song of the time. Yet sightings recorded by new white residents displayed striking consistency with Indigenous descriptions. Witnesses did not speak of a long-necked plesiosaur (the "Nessie" model), but of a serpent-like animal, between forty and fifty feet long, dark green or black in color, capable of moving at considerable speed while creating a series of humps breaking the water's surface.

The event that cemented Ogopogo's reputation as a genuine phenomenon occurred in 1926—seven years before the Loch Ness boom. Approximately thirty carloads of people, stopped along the beach at Okanagan Mission, simultaneously observed a massive creature disturbing the waters. It was not a solitary hallucination, but a mass sighting that forced local authorities to consider the biological hypothesis.

The most debated visual evidence emerged decades later, in 1968, through footage filmed by Art Folden. The film shows a dark object moving against both current and wind, generating a wake incompatible with that of a boat or drifting log. Analysis suggests an animal swimming with vertical undulating motion—characteristic of marine mammals rather than fish or reptiles. This detail led some cryptozoologists, including Roy Mackal, to propose that Ogopogo might be a surviving Basilosaurus, a primitive, serpent-like archaeocete whale officially considered extinct since the Eocene epoch.

Although the notion of a prehistoric cetacean inhabiting a closed freshwater lake is ecologically improbable, the morphology described by witnesses—a horse- or sheep-like head, elongated body, smooth skin—aligns surprisingly well with fossil reconstructions of these ancient marine predators.

Skeptics, supported by limnologists, offer physical explanations tied to the peculiar nature of Lake Okanagan. As a long, narrow, and deep basin, it is prone to the phenomenon of seiches—internal standing waves that can produce anomalous, seemingly self-propelled surface disturbances. Other theories point to optical illusions created by groups of otters swimming in single file, simulating the back of a single enormous serpent. Despite rational explanations, Ogopogo remains legally protected in Canada. In 1989, it was symbolically added to British Columbia's list of protected species—an unparalleled legal acknowledgment that affirms the cultural, if not biological, significance of the great serpent of the lake.

Okanagan-Similkameen, Columbia, V0H 1Z6, Canada

41°40'44"N 82°50'57"W

Ogopogo 1968 video

Ogopogo Videos and Documentaries

Ogopogo – Film

Artificially generated image for illustrative and narrative purposes

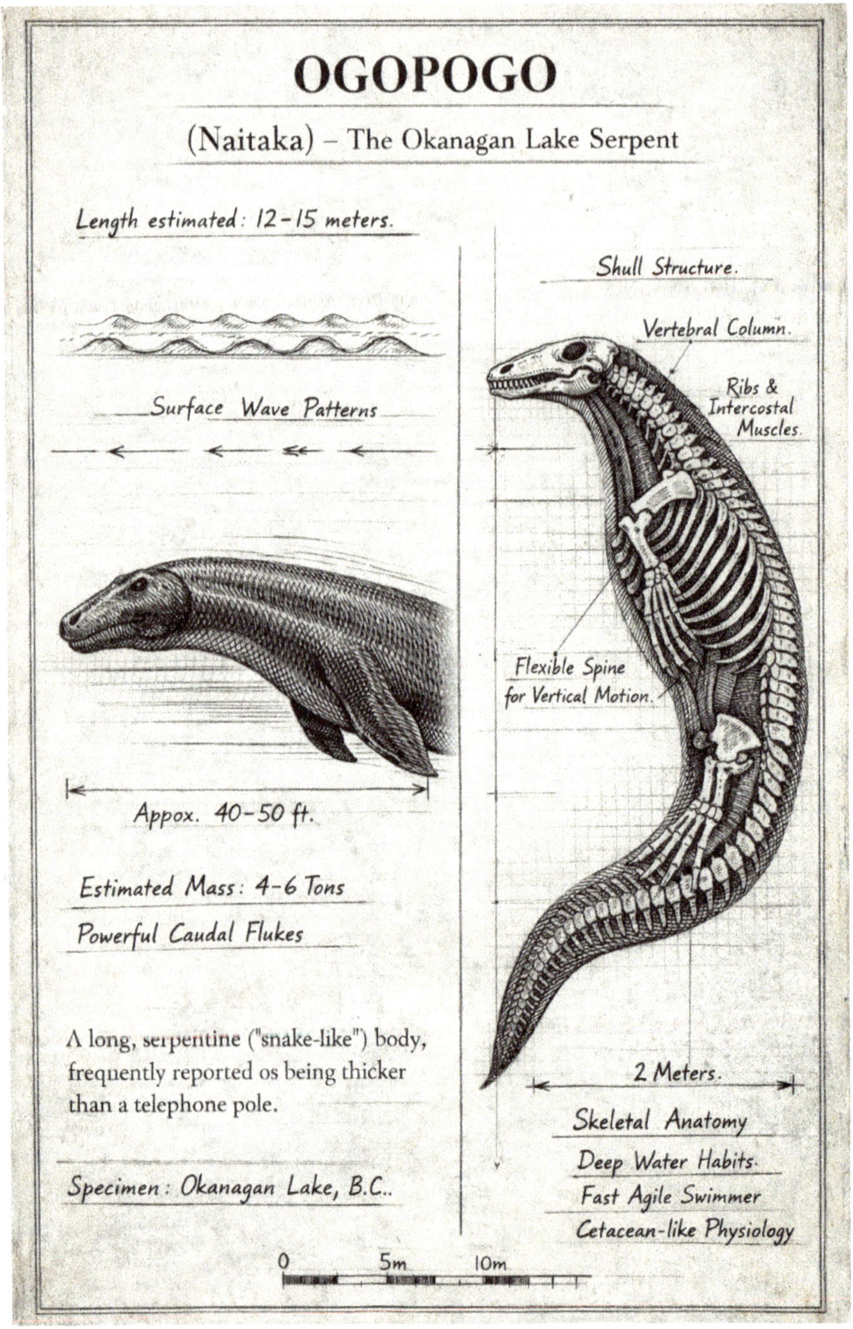

Artificially generated image for illustrative and narrative purposes

Lariosaurus

In the landscape of global cryptozoology—dominated by oceanic giants and Scottish monsters—Italy lays claim to its own peculiar anomaly in the deep waters of Lake Como. The case of the Lariosaurus is a textbook example of how an incontrovertible scientific fact can, through the game of telephone of local reporting, transform into a contemporary legend. Unlike Nessie or Bigfoot, whose biological existence remains purely hypothetical, the Lariosaurus is real. Or at least, it was 235 million years ago.

In 1830, the paleontologist Giuseppe Balsamo Crivelli identified in the limestone sediments of Perledo, on the Lecco shore of the lake, the fossil remains of a Middle Triassic aquatic reptile, which he named Lariosaurus balsami. It was an amphibious predator belonging to the nothosaur group, measuring between two and three feet in length, equipped with webbed limbs and an elongated neck, hunting in the coastal waters of the ancient Tethys Sea when the Alps had not yet risen from the ocean.

The modern "monster," however, was born much later—precisely in November 1946—in an Italy seeking to leave behind the horrors of war, even through the distraction of more harmless mysteries. Two hunters from Como reported sighting, near the Pian di Spagna Nature Reserve at the northern end of the lake, a creature several yards long, with reddish, scaly skin, that slipped hissing into the reeds. Local newspapers, hastily retrieving the scientific name of the regional fossil, headlined the appearance of the "Lariosaurus." From that moment on, a persistent zoological misunderstanding took root: the small, one-meter fossil reptile became, in popular imagination, a ten- or fifteen-meter beast capable of capsizing boats.

Sightings continued sporadically for decades, fueled by the very nature of the Lario itself. With a maximum depth of 410 meters (in the Argegno basin), Lake Como is the deepest lake in Italy and one of the darkest in Europe, riddled with underwater caves and sudden currents. In 1954, news reports described a rounded-snouted creature with webbed limbs sighted near Argegno. In 1957, biologist Giorgio Cappelletti, descending in a homemade bathysphere, claimed to have seen a monstrous shadow brush against his hull in the dark depths. More recently, in 2003, international headlines proclaimed the discovery of a gigantic eel over thirty feet long, filmed during a geological survey; it later turned out to be a school of fish swimming in formation—a classic sonar optical illusion.

What truly lurks in the Lario? Science offers candidates less exotic than dinosaurs, yet equally impressive. The lake hosts sturgeon (Acipenser naccarii), which can grow to considerable size and possess an armored appearance reminiscent of prehistoric reptiles. Moreover, the invasive introduction of the wels catfish (Silurus glanis), a voracious predator that easily exceeds eight feet in length, could account for many modern sightings of "long, dark shapes" rippling the surface.

Yet the enduring fascination of the Lariosaurus lies in its bond with stone. The monster is a spectral projection of the fossils sleeping in the surrounding mountains—a geological ghost that the lake's restless currents seem eager to revive, transforming every anomalous wave into the back of an ancient master returned to reclaim its waters.

Fossa di Argegno, Como, Italy

45°56'40"N 9°08'25"E

Lariosaurus Videos and Documentaries

The Monster of Lake Como Videos and Documentaries

Museo di Storia Naturale di Lecco

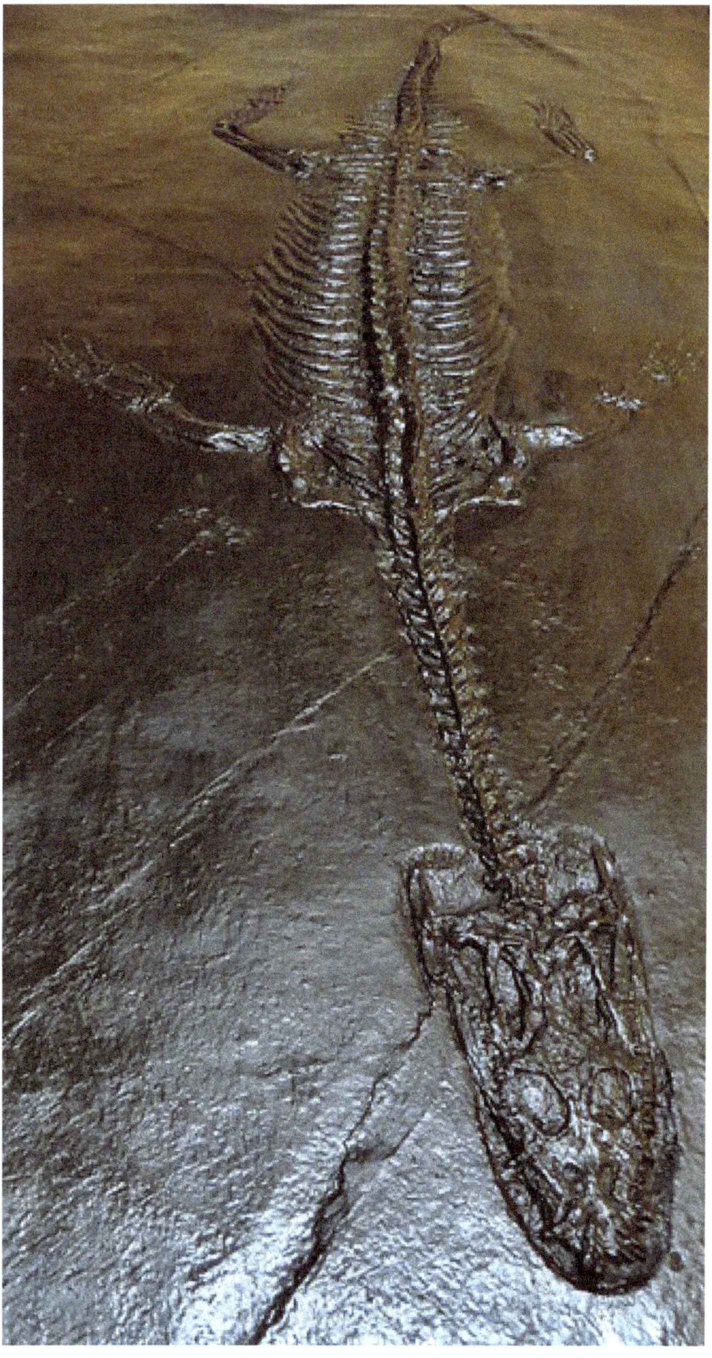

Fossil of Lariosaurus balsami

Artificially generated image for illustrative and narrative purposes

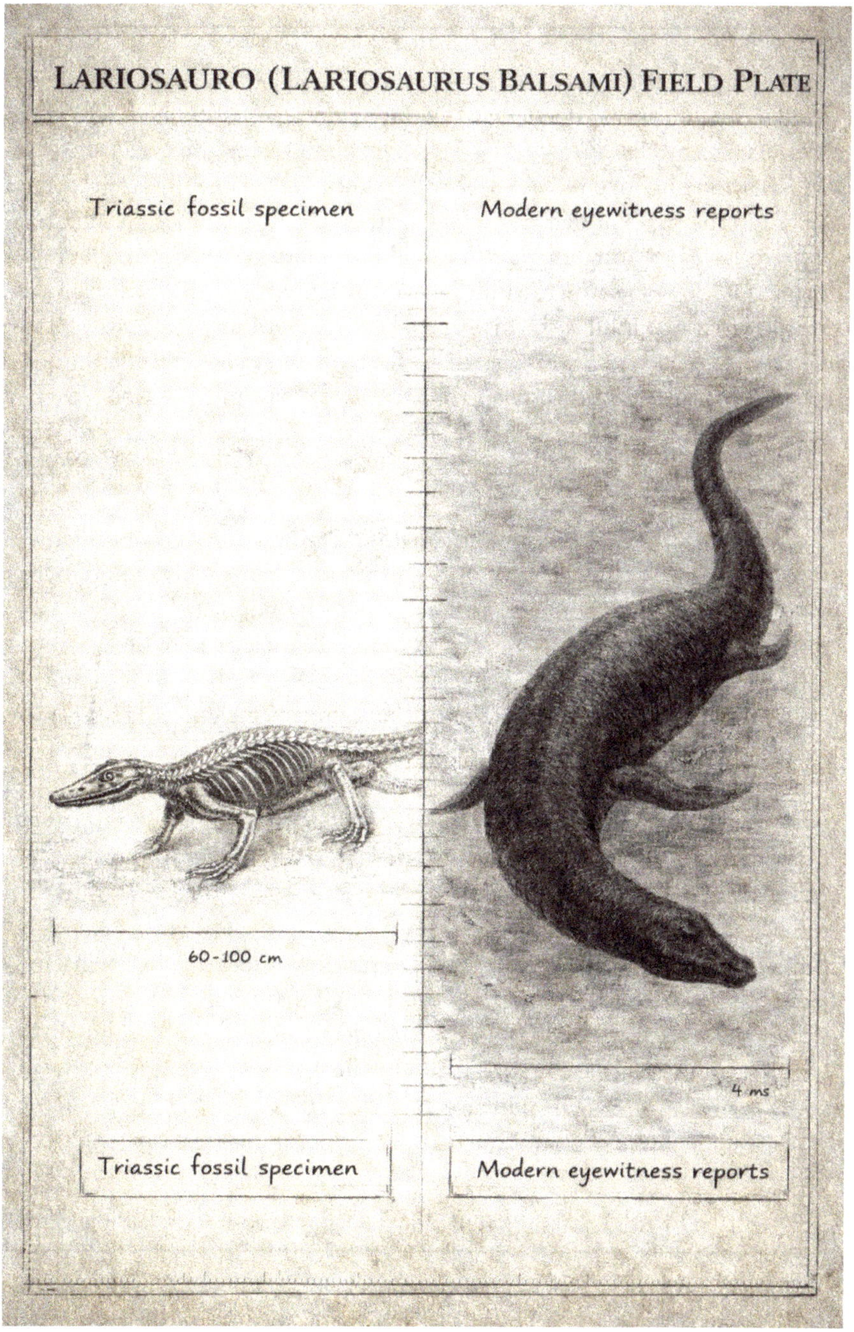

Artificially generated image for illustrative and narrative purposes

Kraken & Globster

Of all the monsters that once haunted ancient nautical charts—often accompanied by the warning Hic Sunt Dracones ("Here Be Dragons")—the Kraken is the only one to have successfully completed the full transition from myth into the pages of marine biology textbooks. For centuries, Scandinavian sailors told of floating islands that suddenly plunged beneath the waves, dragging entire ships down with them, or of tentacles as thick as mainmasts capable of crushing hulls. In his Natural History of Norway (1752), Bishop Erik Pontoppidan described the Kraken as the largest creature in creation, so immense that it could be mistaken for a stretch of coastline. It seemed like pure Nordic folklore—until the nineteenth century, when the sea began to return physical evidence.

The pivotal episode occurred in 1861, when the French corvette Alecton, sailing off the coast of Tenerife, engaged in a surreal battle with a squid roughly twenty-three feet long. The crew managed to harpoon it, but the creature escaped, leaving behind only a fragment of its tail. That piece of flesh, analyzed by the French Academy of Sciences, marked the end of legend and the beginning of teuthology—the scientific study of cephalopods. The monster had a name: Architeuthis dux, the Giant Squid.

Today we know these invertebrates can reach lengths of over forty feet, possess eyes as large as basketballs to detect bioluminescence in the abyss, and have three hearts pumping copper-based blue blood. Yet the aura of mystery endured well into the twenty-first century. For nearly 150 years, science had studied only carcasses washed ashore or remains found in the stomachs of sperm whales. Only in 2004—and later, in high-definition footage in 2012—did Japanese researcher Tsunemi Kubodera succeed in filming a live Architeuthis in its natural habitat. The images revealed not a sluggish, gelatinous beast, as long imagined, but a swift, metallic predator. The Kraken exists—and reality is even more alien than medieval imagination.

But if the Kraken has found its explanation, the ocean continues to cast ashore biological enigmas that defy immediate identification: the so-called "globsters." The term, coined in 1962 by Ivan T. Sanderson to describe a shapeless organic mass stranded in Tasmania, refers to those gelatinous remains—lacking visible bones, eyes, or heads—that appear to belong to enormous unknown octopuses or even extraterrestrial lifeforms.

The most famous case remains the "St. Augustine Monster," which washed up in Florida in 1896: a massive heap of white, fibrous tissue weighing several tons that early analyses claimed belonged to a gigantic octopus (Octopus giganteus) far larger than the Giant Squid. For a century, cryptozoologists and biologists debated the nature of such findings, which periodically surfaced from the coasts of Chile to the Hebrides. The mystery of the globsters was resolved only with the advent of DNA analysis and electron microscopy, revealing a decomposition process as fascinating as it is macabre. Nearly all the celebrated "gelatinous monsters" of the past—including the St. Augustine specimen and the 2003 "Chilean Blob"—turned out to be enormous masses of pure collagen belonging to whales. When a cetacean dies and drifts, its fat and skin decompose, and its heavy bones detach and sink into the abyss. What remains is the tough, fibrous subcutaneous connective tissue sac, shaped by ocean currents into vaguely tentacled or amorphous forms. The "monster" is, in truth, the structural ghost of a whale, stripped by the sea until it becomes an unrecognizable enigma to the human eye.

St. Augustine Beach, Florida, USA

29°50'43"N 81°16'02"W

Mostro di St. Augustine Videos and Documentaries

Architeuthis, 2012 Videos and Documentaries

Globster Videos and Documentaries

Giant squid specimen stranded in Trondheim, Norway, 1954

Artificially generated image for illustrative and narrative purposes

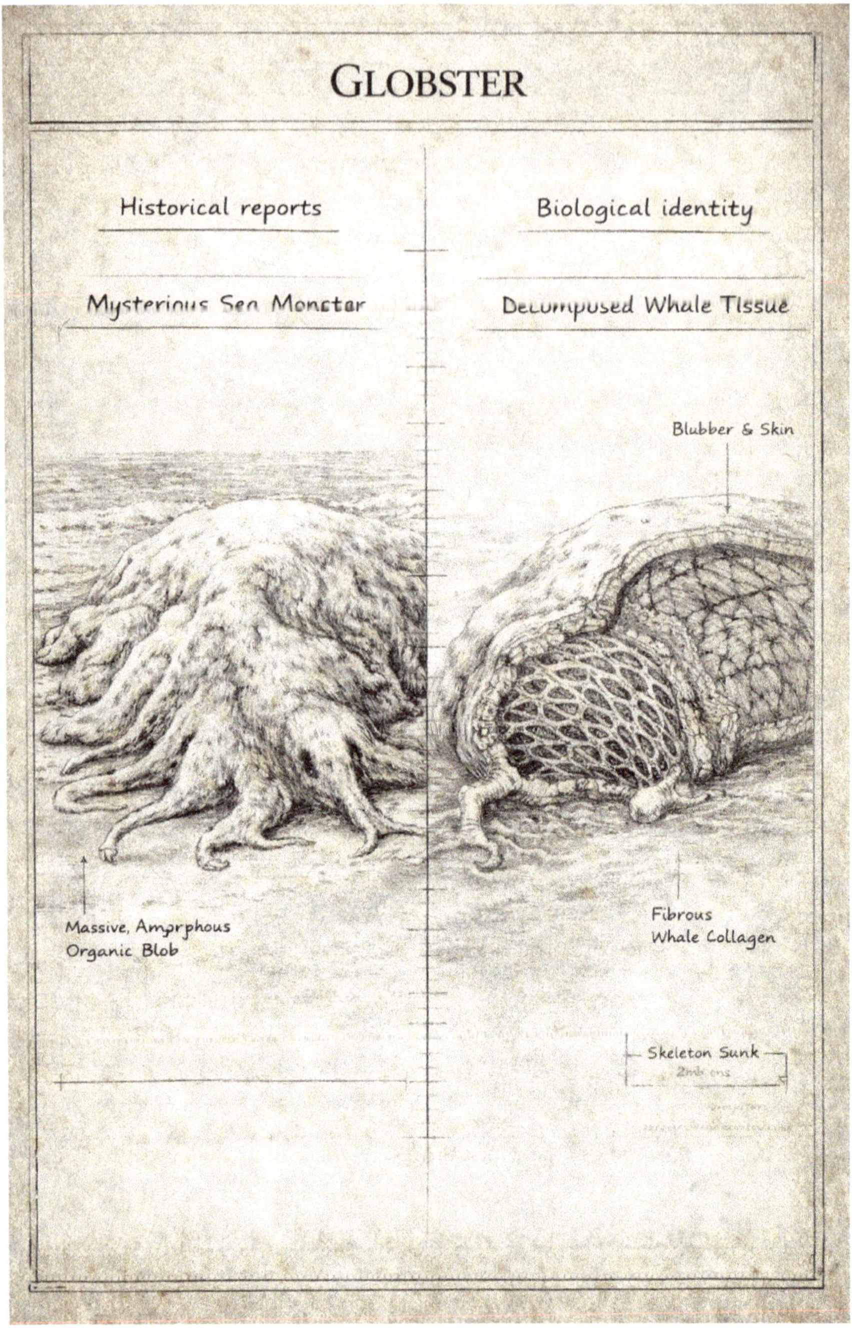

Artificially generated image for illustrative and narrative purposes

Mokele-Mbembe

If Loch Ness is the bourgeois drawing room of cryptozoology—accessible and touristic—the Congo River basin is its true heart of darkness, the last inaccessible frontier where Arthur Conan Doyle's dream of a "Lost World" might still have a biological foundation. In a vast expanse of swamps and equatorial rainforest roughly the size of Florida, almost entirely uninhabited and known as the Likouala region, the local Aka and Fang pygmy tribes have for generations described a creature that official zoology considers extinct for sixty-five million years. Its local name, Mokele-Mbembe, means "the one who stops the flow of rivers," a title suggesting colossal size, capable of altering the very hydrology of the land.

The West first learned of the myth in 1913, when German Captain Freiherr von Stein zu Lausnitz, engaged in colonial mapping in Cameroon, collected accounts of an animal "with smooth grayish-brown skin, as large as an elephant but with a long, flexible neck and a single tooth—or perhaps a very long horn." The description inevitably evoked a small sauropod, a herbivorous dinosaur similar to Apatosaurus, adapted to amphibious life. For decades, this image fueled adventurous expeditions searching the Congo for a climatic refuge where prehistory had never truly ended.

In the 1980s, University of Chicago biologist Roy Mackal led two missions into Likouala, gathering remarkably consistent testimony. When shown illustrated zoology manuals without prompting, local informants ignored images of elephants, hippos, and bears, but decisively pointed to the sauropod illustration, stating: "That is Mokele-Mbembe."

The creature is described as strictly herbivorous, fond of the fruit of a local riparian plant, Landolphia, yet highly territorial and aggressive—capable of overturning canoes and killing humans without consuming them. Despite the narrative consistency, physical evidence remains nonexistent. The acidity of jungle soil rapidly dissolves bone, preventing recent fossilization, and the density of the forest canopy renders satellite observation useless. The few "stolen" images—such as a brief Japanese film from 1992 showing a wake and a dark object in Lake Tele—are inconclusive.

Skeptics propose real-world candidates to explain the monster. The strongest hypothesis points to the black rhinoceros, once present in the region and now locally extinct: a massive, dangerous animal with a horn that, when submerged up to its neck, could deceive the eye. Another possibility is the African softshell turtle (Trionyx triunguis), which can reach impressive size; its long neck and small head emerging from the water, seen in profile, resemble the silhouette of a "mini-dinosaur."

Yet Mokele-Mbembe endures as the last great zoological dream of the nineteenth century: the idea that somewhere on Earth there remains a corner so remote and hostile that a dragon could have survived the asteroid that killed its kin—not protected by magic, but by the impenetrability of the rainforest.

Lake Tele, Congo

1°20'52"N 17°09'16"E

Mokele-Mbembe Videos and Documentaries

Spedizione di Roy Mackal (anni '80) Videos and Documentaries

Artificially generated image for illustrative and narrative purposes

Artificially generated image for illustrative and narrative purposes

SECTION III.
Aerial and Atmospheric Phenomena

Mothman

No case in the history of twentieth-century strangeness intertwines collective hysteria, UFO investigation, and civil tragedy as precisely as that of the Mothman. If Bigfoot belongs to the wilderness and Nessie to lost geological ages, Mothman is a distinctly modern, industrial, and peri-urban monster. His story is confined to a tightly defined span of space and time: exactly thirteen months, from November 1966 to December 1967, in the town of Point Pleasant, West Virginia. This small community along the Ohio River, surrounded by forests and abandoned chemical facilities, became the stage for a "window of strangeness" that drew national media attention and investigators such as John Keel, whose chronicles immortalized the legend.

It began on the night of November 15, 1966, when two young couples—the Scarberrys and the Mallettes—were driving near a former World War II munitions plant known locally as the "TNT Area." Isolated and dotted with vegetation-covered concrete bunkers, the site was the perfect setting for a terrifying encounter. The car's headlights illuminated two large, glowing red circles, described by witnesses not as mechanical lights but as living eyes. They belonged to a gray figure about seven feet tall, vaguely humanoid but lacking a visible neck: its head appeared to rest directly on its shoulders—or to be absent altogether—with the eyes positioned on its chest.

When the creature unfurled a pair of enormous wings, like those of a bat or a giant moth, and rose vertically into the air, panic set in. The most disturbing detail reported to Sheriff George Johnson was that the creature pursued the car, which was traveling at nearly 100 miles per hour, gliding above it without flapping its wings and emitting a high-pitched sound likened to a large rodent's squeal or a record spinning at extreme speed.

In the months that followed, Point Pleasant was flooded with more than one hundred similar sightings. Firefighters, pilots, and housewives reported seeing the "giant bird" perched on trees, rooftops, and power lines. The phenomenon extended beyond cryptozoology: it was accompanied by alleged electromagnetic interference affecting radios and televisions, reports of animal mutilations, and encounters with mysterious "Men in Black," creating a climate of widespread paranoia that Keel described as "high strangeness."

Biology, however, offers rational explanations for this winged nightmare. Ornithologists have frequently pointed to the Sandhill Crane (Antigone canadensis), a bird that can stand nearly as tall as a human, with a wingspan of over six feet and, crucially, a patch of bare red skin on its head that can appear strikingly vivid under stress or certain lighting conditions. Another candidate is the Great Horned Owl (Bubo virginianus), the largest nocturnal raptor in North America: its eyes reflect headlights with an intense orange-red glow (eyeshine), and its flight is virtually silent. The "lack of a neck" described by witnesses is consistent with the stocky silhouette of a perched owl.

Yet the legend of Mothman feeds not on ornithology but on fatality. The creature entered folklore not as an animal but as a harbinger of disaster. The wave of sightings ended abruptly and tragically on December 15, 1967, at 5:04 p.m. The Silver Bridge, which connected Point Pleasant to Ohio, suddenly collapsed during rush hour due to the structural failure of a single eyebar in its suspension chain. Forty-six people died in the icy waters of the river.

In the collective psyche, the red-eyed monster—seen perched on the bridge's metal structures in the days before the collapse—became retroactively the herald of catastrophe, perhaps even its supernatural cause. Engineering science has since explained the disaster through corrosion and metal fatigue, but the myth endures: Mothman remains the archetype of the "prophetic monster," a projection of our anxieties onto disasters we cannot foresee or control.

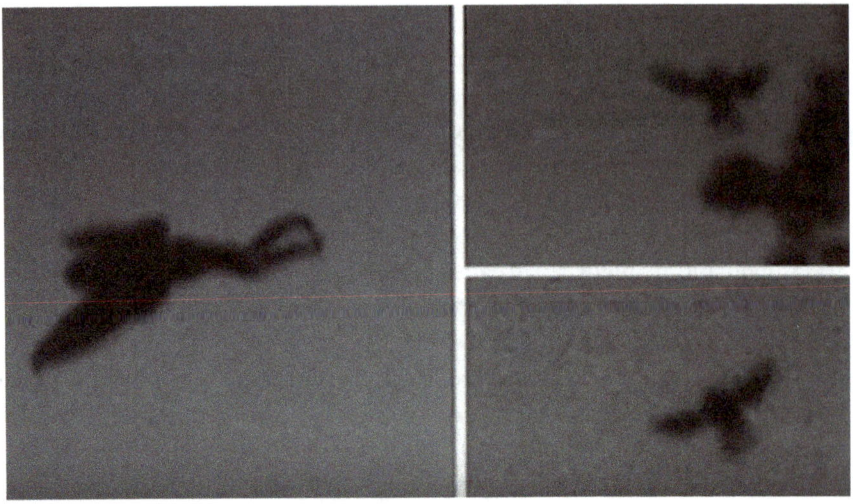

Foto del 2016 – Fonte: WCHSTV

6182 Ohio River Rd, Point Pleasant, WV 25550, USA	38°55'46"N 82°04'33"W	Interviews with Mothman witnesses, 1966, Point Pleasant
Videos and Documentaries	Mothman Museum	Mothman – Film

Artificially generated image for illustrative and narrative purposes

Artificially generated image for illustrative and narrative purposes

Thunderbird

If Mothman embodies modern, industrial horror, the Thunderbird carries us back to ancestral fears—when humans were not the dominant predator but potential prey, scanning the sky with apprehension. In Native American cultures, from the Algonquian peoples of the East Coast to the Sioux of the Great Plains, the Thunderbird is a powerful deity: a spirit that brings rain and hurls lightning from its eyes. For cryptozoology, however, this entity is not confined to the spiritual realm. A distinct line of inquiry explores the possibility that North America still harbors—or may have harbored until very recent times—a species of colossal raptor, biologically incompatible with any known avian fauna.

The most dramatic and well-documented case, which transformed myth into headline news, occurred on the evening of July 25, 1977, in Lawndale, central Illinois. Three boys were playing in their backyard when two enormous birds—described as condor-like, with an estimated wingspan exceeding ten feet—swooped down. One of the birds reportedly seized ten-year-old Marlon Lowe, who weighed about sixty-five pounds, by the shoulders, lifting him more than a foot off the ground and carrying him nearly thirty feet toward a wooded area before his mother's screams forced the predator to release him.

The Lawndale incident was not isolated. Nineteenth-century pioneer and gold prospector accounts are filled with stories of "monstrous eagles" capable of carrying off lambs and even calves. Faced with such reports, science stands at a crossroads. On one side lies the hypothesis of surviving Teratorns (family Teratornithidae). These prehistoric birds—such as Teratornis merriami or the gigantic Argentavis magnificens, whose wingspan reached over twenty-three feet—coexisted with early humans during the Pleistocene. Officially extinct for ten thousand years, their morphology aligns strikingly well with Thunderbird descriptions: enormous bodies, hooked beaks, and powerful talons.

Could a relict population have survived in the most remote regions of the Rocky Mountains or Alaska? Skeptical ornithologists counter with the square-cube law and energy requirements: a predator of that scale would need vast hunting territory and specific thermal updrafts to sustain flight, making it nearly impossible to remain undetected in the age of radar and satellites. A far more plausible explanation, according to mainstream science, is the misidentification of California Condors or Golden Eagles, whose impressive dimensions can be grossly exaggerated by frightened witnesses or distorted by deceptive perspective.

A second mystery surrounding the Thunderbird is psychological and media-driven: the "Lost Photo." Thousands of people—including seasoned researchers—swear they once saw, in old schoolbooks or 1950s pulp magazines, a black-and-white Wild West–style photograph depicting a gigantic dead bird nailed to a barn wall, with a row of men posing in front of it, arms outstretched to demonstrate its scale. Despite decades of archival searches, no such image has ever been found. Today, the photo is widely considered one of the most striking examples of the "Mandela Effect"—a shared false memory. The human mind likely stitched together written descriptions and unrelated images (such as film depictions of pterodactyls), creating a vivid but nonexistent visual memory. The Thunderbird thus soars on two planes: the physical one, where a boy in Illinois still bears unexplained talon scars, and the realm of memory, where the image of a monstrous bird pinned to a barn continues to exist only in our minds.

Lawndale, Chicago, Illinois, USA

41°51'35"N 87°42'54"W

Thunderbird, 1977 Videos and Documentaries

The Mystery of the Thunderbird Videos and Documentaries

The Mystery of the Thunderbird and the Mandela Effect Videos and Documentaries

Artificially generated image for illustrative and narrative purposes

Artificially generated image for illustrative and narrative purposes

Jersey Devil

In the heart of the American East Coast, wedged between the frenetic metropolises of New York and Philadelphia, lies a geographic and ecological anomaly: the New Jersey Pine Barrens. This vast expanse of stunted pine forest, acidic swamps, and sandy soil covers more than 1,700 square miles and has, since colonial times, been a landscape of isolation and social marginality. It is within this distinctly American Gothic setting that the Jersey Devil resides—a creature that holds a unique distinction: it is perhaps the only cryptid whose origins can be traced with remarkable precision to a documented blend of eighteenth-century political satire, religious folklore, and industrial-era mass hysteria.

The canonical legend fixes the monster's birth in 1735, on a stormy night in Leeds Point. Mother Leeds, an impoverished woman already burdened with twelve children, allegedly cursed her thirteenth pregnancy, crying out, "Let it be the devil!" According to the tale, the child was born seemingly normal, only to transform within moments into a grotesque chimera—with a horse's head, bat-like wings, cloven hooves, and a forked tail—before killing the midwife and escaping up the chimney into the surrounding woods.

Modern historians, however, including Brian Regal, have uncovered a less supernatural but historically compelling explanation. The Leeds family was real, as was their rivalry with local Quakers and, notably, with Benjamin Franklin. Daniel Leeds and his son Titan published almanacs that competed directly with Franklin's Poor Richard's Almanack. In the fierce pamphlet wars of the era, Franklin publicly mocked the Leeds family, referring to them as "ghosts" and "monsters." The Leeds family crest itself featured dragon-like wyverns. It is highly likely that the "Leeds Monster" began as political caricature—a means of discrediting a rival family—gradually morphing into a literal beast through decades of superstition among the isolated inhabitants of the Barrens, known as "Pineys."

If the origins were political, the endurance of the phenomenon is biological and social. The tipping point occurred between January 16 and 23, 1909, during what came to be known as the "Phenomenal Week." Over the course of seven days, thousands of people throughout the Delaware Valley reported sightings of the creature. Schools were closed, factories halted production out of fear, and police in Camden and Philadelphia fired at moving shadows. The composite sketches collected at the time were strikingly consistent, describing a bipedal creature approximately six feet tall, with a kangaroo-like body, membranous wings, and an equine head.

Zoology struggles to identify a local species that matches this morphology. The only real animal that closely resembles the Jersey Devil—with its elongated horse-like face and large wings—is the Hammer-headed Bat (Hypsignathus monstrosus), a species native to equatorial Africa and entirely incapable of surviving a New Jersey winter. Skeptics instead point to the Sandhill Crane (often mistaken for other large cranes), which during mating displays may leap and spread its wings in ways that, in fog or low light, could appear monstrous. Another, more somber hypothesis suggests that the original 1735 legend may have stemmed from the birth of a severely deformed child in the Leeds family—hidden from view and dehumanized by rumor.

Today, the Jersey Devil has been domesticated into state folklore and even lends its name to a professional hockey team. Yet for those who wander at night through the Pine Barrens—among the Atlantic white cedars and the dark waters of the Mullica River—the winged silhouette remains something less playful: a shadow from an archaic America that modernity has never quite managed to clear away.

Leeds Point, Galloway,
New Jersey 08205
USA

39°29'34"N 74°26'34"W

The Jersey Devil, 1909
Videos and
Documentaries

The Jersey Devil
Videos and
Documentaries

The Jersey Devil
"Sounds"

The Jersey Devil
Film

Japhet Leeds House, Moss Mill Road, Leeds Point, New Jersey, 1937 ca.

JERSEY TERROR IN TOWN; NIGHT WATCHMAN SAW IT

Hovered Over Southern Part of the City Near the Paper Mill.

watchman at the

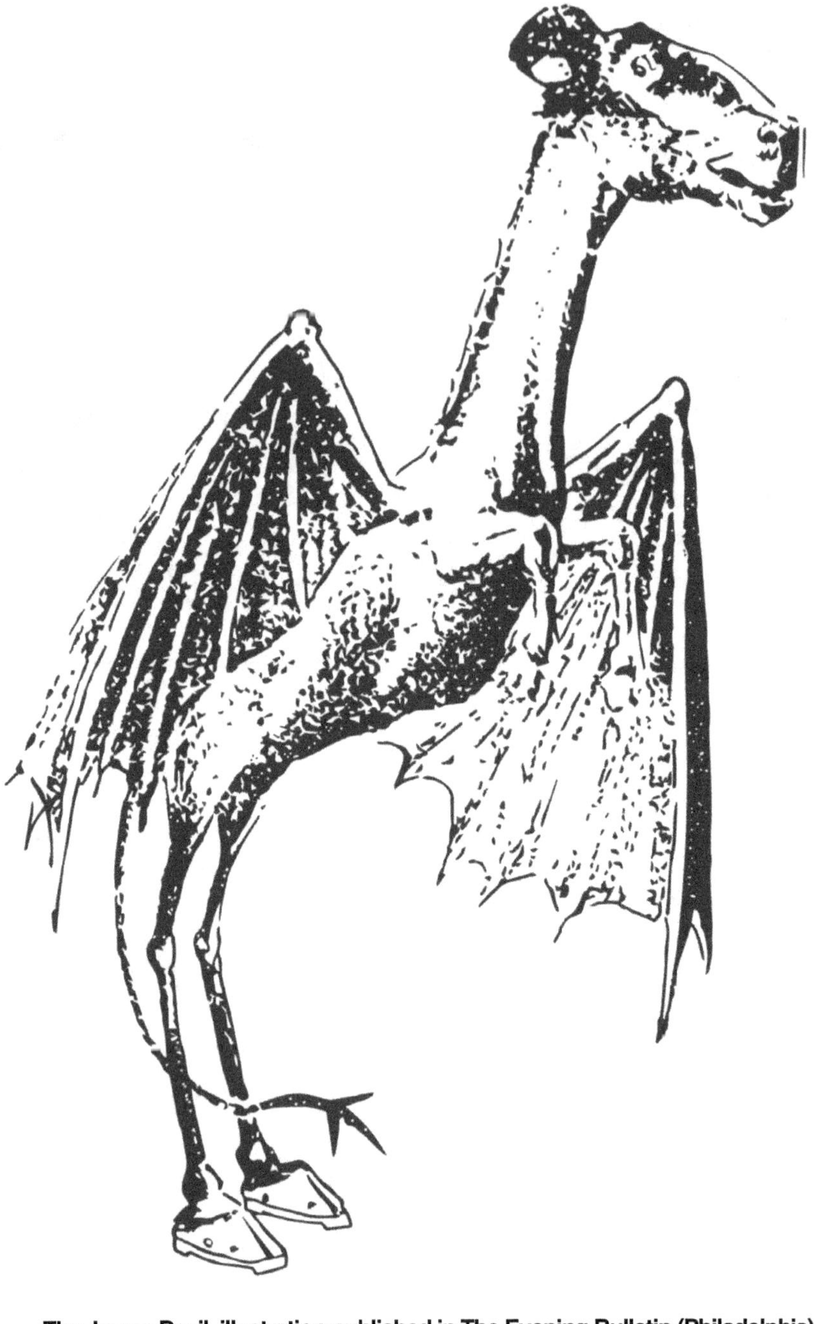

The Jersey Devil, illustration published in The Evening Bulletin (Philadelphia), January 21, 1909

Artificially generated image for illustrative and narrative purposes

Artificially generated image for illustrative and narrative purposes

SECTION IV.
Speculative Biology and Hybrids

Chupacabra

If Bigfoot is the monster of analog folklore and word-of-mouth storytelling, the Chupacabra holds the historic distinction of being the first monster of the Internet age. Its origin is remarkably recent and documented with a precision that most other cryptids lack. We do not need to look back to dark centuries or ancient tribal legends: the "goat-sucker" was officially born in March 1995 in Canóvanas, a rural municipality in Puerto Rico. During that period, farmers began finding goats, sheep, and domestic animals dead, apparently drained of blood through small circular puncture wounds in the neck. Puerto Rican comedian and entrepreneur Silverio Pérez, commenting on the news on the radio, coined the term Chupacabras almost mockingly, but the name immediately achieved viral resonance, spreading through early chat rooms and online forums, riding the wave of sensationalist Spanish-language television programs such as Ocurrió Así.

What makes the Chupacabra case unique from both a sociological and zoological perspective is its radical physical metamorphosis within the span of a single decade. In fact, there are two distinct monsters. The "Original Chupacabra" of 1995, described by key witness Madelyne Tolentino, was a bipedal being about four feet tall, with large elongated red eyes, thin arms, and a crest of spines running down its back, resembling a reptilian alien or a space gargoyle. This description fueled years of UFO-related theories, linking the animal mutilations to alleged genetic experiments conducted by the United States in the nearby El Yunque rainforest. However, in 2011, researcher Benjamin Radford conducted a revealing comparative investigation: Tolentino's description matched almost perfectly the design of the alien creature "Sil" from the sci-fi horror film Species, which had premiered in Puerto Rico just weeks before the first sightings. It is a textbook case of cultural contamination: the recent and terrifying cinematic image had provided the witness's brain with the "template" to process a traumatic or confusing experience.

Around the year 2000, however, the monster shed its skin—and its habitat. Moving from Puerto Rico to Mexico and eventually to Texas, the Chupacabra ceased to be a bipedal alien and became a quadruped. The "Texas Chupacabra," the one that appears today in viral photos and border patrol reports, is described as a hairless canid with grayish or bluish skin, protruding teeth, and a skeletal appearance. Unlike the Puerto Rican version, here cryptozoology gives way to veterinary medicine. Every captured or deceased specimen subjected to DNA testing has turned out to be a common coyote (Canis latrans), domestic dog, or raccoon suffering from a severe form of sarcoptic mange.

Mange, caused by the mite Sarcoptes scabiei, transforms a familiar animal into something monstrous and unrecognizable. The complete loss of fur exposes the skin, which thickens and takes on a bluish or grayish, elephant-like hue; the animal, weakened and feverish, can no longer hunt swift wild prey such as rabbits and is forced to approach human settlements to attack slower, easier domestic livestock. Even the myth of "vampirism" finds a forensic explanation: a sick and debilitated canid often kills its prey but lacks the strength to tear it apart or consume it fully; as blood coagulates rapidly inside an unopened carcass, it creates the illusion that the animal has been "drained." The Chupacabra, then, does exist—but it is not an alien predator. It is the tragic and deformed face of animal disease, transformed into a nightmare by our inability to recognize what is familiar once it has been stripped of its "fur coat" of normality.

Canóvanas, 00729
Puerto Rico

18°22'50"N 65°54'07"W

Chupacabra
Videos and
Documentaries

Chupacabra
Interviews and
Discussions

Chupacabra
Film

Artificially generated image for illustrative and narrative purposes

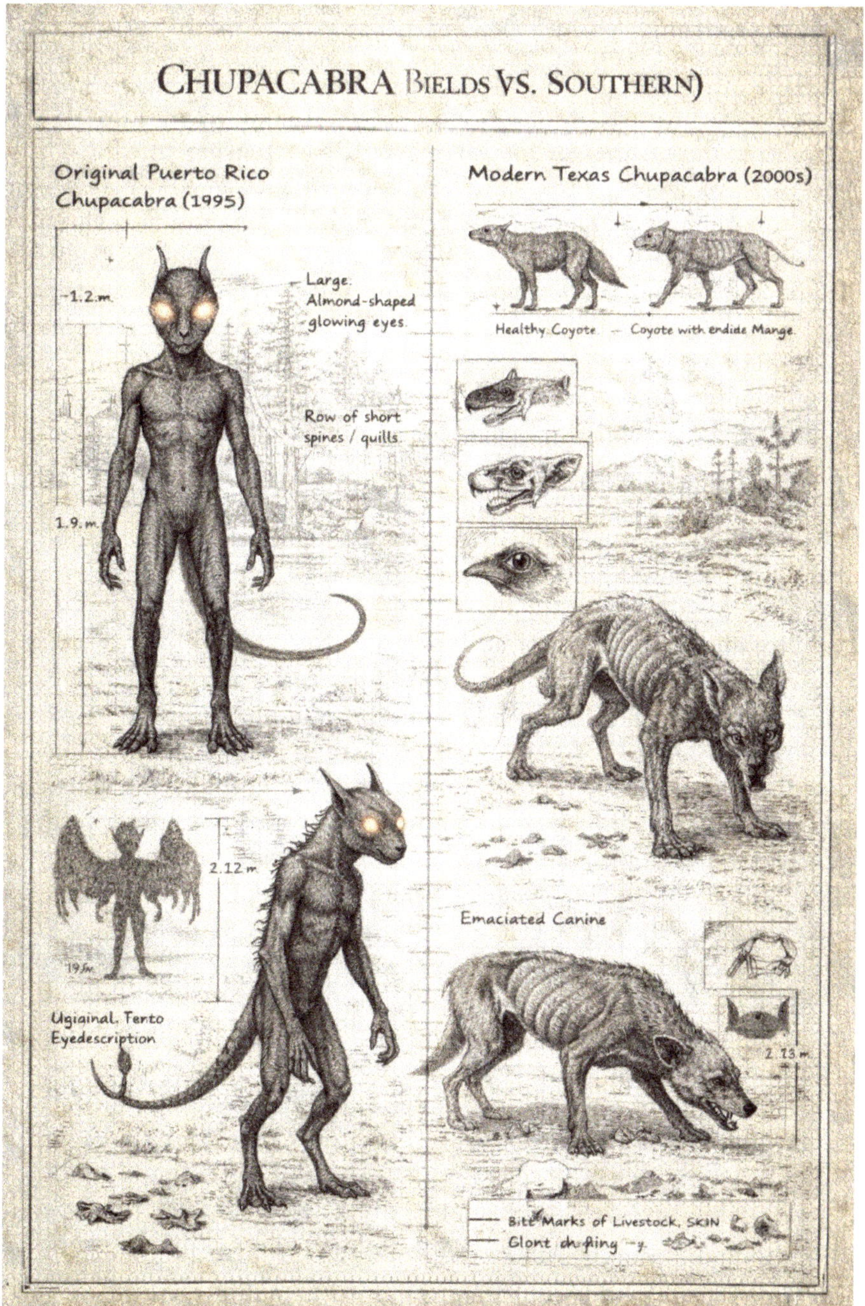

Artificially generated image for illustrative and narrative purposes

Tatzelwurm

In the collective imagination, the Alps are the realm of ibex, marmots, and eagles. Yet the archives of German- and Italian-speaking naturalists have preserved, for at least three centuries, traces of a creature that defies the taxonomy of European reptiles. The Tatzelwurm—whose name literally means "worm with paws" (also known as Stollenwurm in Switzerland and Basilisk in Trentino)—does not belong to the aerial mythology of fire-breathing dragons, but to the damp and shadowed world of the subterranean. Descriptions collected between the eighteenth century and the mid-twentieth century from hunters and alpine herders are strikingly consistent—a factor that, in cryptozoology, is often considered indicative of a possible underlying biological reality. The animal is described as a reptile measuring between twenty inches and three feet in length, with a stout, cylindrical body resembling that of a snake, smooth or finely scaled skin, a flattened, vaguely feline head, and—crucially—only two front legs equipped with claws.

What distinguishes the Tatzelwurm from a mere folktale is the attention it received from Enlightenment and Romantic-era scientists. In 1723, Swiss naturalist Johann Jakob Scheuchzer, in his monumental Itinera per Helvetiae, cataloged various types of Alpine "dragons" with an almost Linnaean approach, attempting to separate fact from superstition. But it was in the nineteenth century that the pursuit became more serious. In 1861, an authority such as Friedrich von Tschudi, in Das Tierleben der Alpenwelt (The Animal Life of the Alpine World), mentioned the existence of an unknown reptile called the Stollenwurm, suggesting it might be an unclassified species living in deep rock fissures, emerging only rarely and under particular atmospheric conditions—often before thunderstorms. The creature was feared for its aggressiveness and for its alleged venom, or "pestilential breath," capable of making humans ill—a detail that may allude to a real chemical defense mechanism.

If we were to search for a biological candidate for the Tatzelwurm, nature offers surprising parallels—albeit in distant regions. The description of a "stocky venomous lizard" bears an uncanny resemblance to North American helodermatids such as the Gila monster (Heloderma suspectum): slow, heavy-bodied reptiles with fat-storing tails and venomous bites. Could a related species, having survived the glaciations in thermal microclimates or deep cave systems, have inhabited the Alps until relatively recent times? Another zoological hypothesis points toward skinks or amphisbaenians—reptiles that display various degrees of limb reduction. The European skink (Chalcides

chalcides) has tiny, almost vestigial legs, but it is far too small to match the Alpine monster. Some cryptozoologists instead speculate about the existence of a European giant salamander, similar to the Japanese giant salamander (Andrias japonicus), which can reach five feet in length—an amphibian that might have developed more terrestrial habits and toxic defenses.

The Tatzelwurm file suffered a serious setback in 1934, when a Swiss photographer named Balkin published what he claimed was the first photograph of the creature in the Berliner Illustrierte Zeitung. The image showed a wrinkled, aggressive-looking animal emerging from a burrow. Though initially received with enthusiasm, the photo was soon exposed as a crude hoax, likely created using a fish or a ceramic model, casting discredit on the entire affair. Yet the absence of a genuine photograph does not erase centuries of sightings. Although anecdotal reports have continued into the twenty-first century in remote areas of Trentino and South Tyrol, no recent speleological or zoological expedition has ever uncovered skeletal remains that could definitively confirm the species' existence. Perhaps the Tatzelwurm is not a monster at all, but merely an extraordinarily rare and elusive animal—a relic of the pre-glacial era that chose the heart of the mountain as its refuge from humankind, quietly disappearing just as we began searching for it with cameras.

5743 Krimml, Austria

47°12'28"N 12°10'15"E

Tatzelwurm Videos and Documentaries

Artificially generated image for illustrative and narrative purposes

Artificially generated image for illustrative and narrative purposes

Tsuchinoko

While the West scans its lakes for dinosaurs and its forests for giant apes, Japan nurtures a national obsession with a much smaller—yet no less elusive—creature: the Tsuchinoko. The term, which roughly means "child of the hammer" or "child of the earth," refers to a reptile that appears to defy the biomechanics of serpentine movement. Although Japanese culture is populated by yōkai—spirits and supernatural monsters from folklore—the Tsuchinoko occupies a separate category: that of the UMA (Unidentified Mysterious Animal). To the Japanese, it is not a ghost but a flesh-and-blood creature said to inhabit bamboo groves and mountain streams, the object of organized hunts, multimillion-yen bounties, and even pseudo-scientific studies.

The historical roots of this creature run deep—far older than modern urban legends. Ceramic artifacts from the Jōmon period, dating back thousands of years, depict stout, triangular-bodied serpent-like beings that strikingly resemble the Tsuchinoko's anatomy. The Kojiki, the oldest surviving chronicle of Japanese mythology (compiled in 712 CE), also mentions deities and entities that scholars have linked to this cryptid. The physical description, remarkably consistent over the centuries, portrays a snake measuring between twelve and thirty inches in length, with a triangular head clearly distinct from the neck and, most notably, a disproportionately swollen and stubby midsection—often compared to a beer bottle or rolling pin. Unlike ordinary snakes that move through lateral undulation, the Tsuchinoko is said to contract vertically like a caterpillar or even roll itself along the ground. Folklore further attributes surprising athletic and vocal abilities to it: it may leap up to three feet and, according to more fanciful accounts, emit sounds resembling whistles, chirps, or even snoring.

During the twentieth century—especially amid the occult boom of the 1970s and 1980s—the Tsuchinoko became a mass phenomenon. Several rural municipalities, such as the village of Yoshii in Okayama Prefecture and the city of Itoigawa, attempted to stimulate local tourism by offering extravagant rewards—up to 100 million yen (nearly one million dollars)—for anyone who could capture a live specimen. These organized hunts produced a wealth of anecdotal evidence, but no body to examine. Faced with this taxonomic void, herpetology has attempted to provide explanations grounded in visual misidentification.

The most widely accepted theory is that of "difficult digestion." A common Japanese snake—such as the mamushi (a venomous pit viper) or the Japanese rat snake—after swallowing a large prey item like a toad or small rodent, temporarily assumes a swollen midsection, while its head and tail remain slender. Encountering such a snake in this post-feeding state—immobile and oddly thick—could easily create the illusion of an entirely different species. Another intriguing hypothesis involves introduced exotic species. The blue-tongued skink (Tiliqua), a popular terrarium lizard imported into Japan since the 1970s, has a cylindrical, stocky body, a triangular head, and limbs so short that they disappear in tall grass. An escaped or abandoned specimen would closely match the Tsuchinoko's visual profile, explaining its apparent "alien" quality within local fauna.

Although science tends to close the case as a mixture of pareidolia and misidentification, the Tsuchinoko endures as an icon of rural Japan—a countryside unwilling to surrender completely to urbanization. It remains a small underbrush deity, hiding quite literally beneath the feet of one of the most technologically advanced nations on Earth.

Giappone, 〒701-2435 Okayama, Akaiwa, Niborinaka, 2006	34°52'18"N 134°00'47"E	Tsuchinoko Videos and Documentaries

Artificially generated image for illustrative and narrative purposes

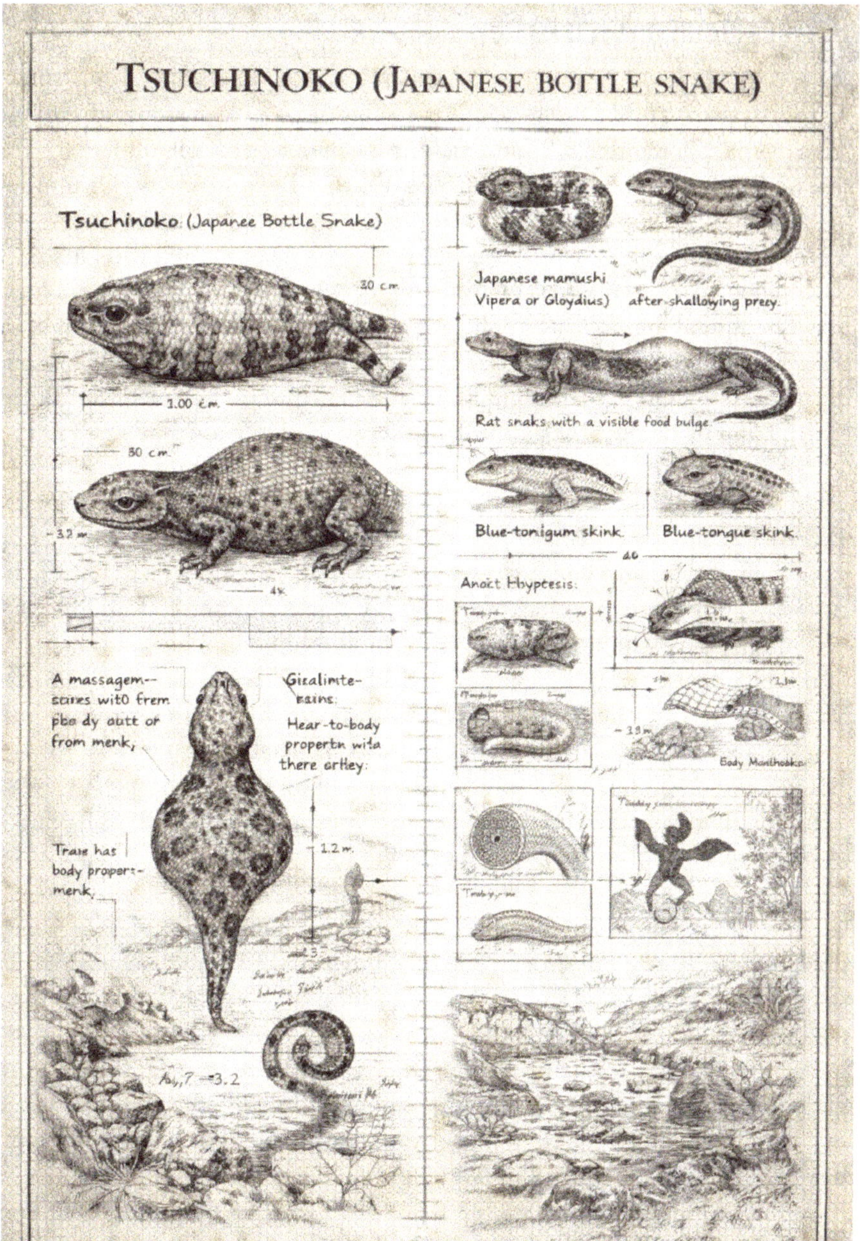

Artificially generated image for illustrative and narrative purposes

Flatwoods Monster

There is a precise point at which zoological inquiry—even its most speculative forms—must come to a halt and yield to something radically different. The case of the Flatwoods Monster, also known as the "Braxton County Phantom," marks that boundary. Here we are not searching for an unclassified ape or a reptile that survived the Cretaceous; instead, we are confronted with an entity that seems to blur the very distinction between biology and technology, between animal and machine. The event, confined to a few frantic hours on the evening of September 12, 1952, in the hills of West Virginia, remains one of the most documented and unsettling close encounters in American history, anticipating by decades the aesthetic of alien visitation narratives.

It began at dusk, when a group of boys playing in the yard of the local elementary school saw a bright object streak across the sky and apparently crash atop a nearby hill on the property of farmer G. Bailey Fisher. Driven by curiosity—and fearing a plane crash—the boys, accompanied by Kathleen May (the mother of two of them) and seventeen-year-old Eugene Lemon, a member of the National Guard, decided to climb the hillside in the dark. What they found was not a burning wreck but an atmosphere thick with a pungent mist that irritated their eyes and throats, described as metallic or sulfurous in odor. At the top of the hill, Lemon spotted two small lights shining through the trees; assuming they were the eyes of an opossum or raccoon, he aimed his flashlight in that direction. The beam revealed something impossible.

The seven witnesses described a towering figure, nearly ten feet tall, hovering just above the ground. The entity possessed a red, ace-of-spades–shaped head or hood, set atop a dark, cylindrical or conical body resembling a pleated metallic "skirt"—perhaps an exoskeleton—with no visible arms except for two small, claw-like appendages. The creature did not walk; it seemed to glide toward them with a shrill hissing sound, while its "face" emitted beams of light piercing the fog. Terror triggered a headlong flight downhill. The most disturbing aspect of the case lies not only in the sighting itself but in the immediate physical aftermath: for weeks, the witnesses reportedly suffered vomiting, convulsions, and throat irritation—symptoms consistent with exposure to blistering agents such as mustard gas, though physicians at the time diagnosed them as traumatic hysteria.

The rational explanation, proposed decades later by investigator Joe Nickell of CSICOP (the Committee for Skeptical Inquiry), attempts to dismantle the monster piece by piece. The object in the sky was almost certainly a meteor—part of a meteor shower documented that night across three states. The creature? Most likely a barn owl (Tyto alba). This nocturnal raptor has a white, heart-shaped facial disk (which in darkness can resemble a spade-shaped head), eyes that reflect light intensely, and a habit of perching on high branches or dead tree trunks. Imagine a barn owl perched atop an old tree, foliage below creating the illusion of a dark "skirt," illuminated from beneath by a trembling flashlight on a foggy night: a frightened human brain could easily connect the dots into a monster. The "hissing" and the apparent "floating" would correspond, respectively, to the owl's defensive screech and its silent downward glide.

Yet the owl theory does not explain everything. It does not account for the witnesses' physical illness, the persistent chemical odor, or the strange oily, rubbery substance reportedly found on the ground the following day by Sheriff Robert Carr and journalist A. Lee Stewart. The Flatwoods Monster endures as an icon because it embodies Cold War anxiety: the threat is no longer a predator that devours you (like Bigfoot), but a technological intruder that contaminates you. Whether it was a poorly lit owl or a malfunctioning extraterrestrial probe, that night in 1952 permanently altered American folklore, introducing the idea that the woods may hold not only animals—but visitors.

Flatwoods, West Virginia, USA

38°42'40"N 80°39'26"W

Flatwoods Monster Videos and Documentaries

Project Blue Book Flatwoods case files

Flatwoods Monster Film

Artificially generated image for illustrative and narrative purposes

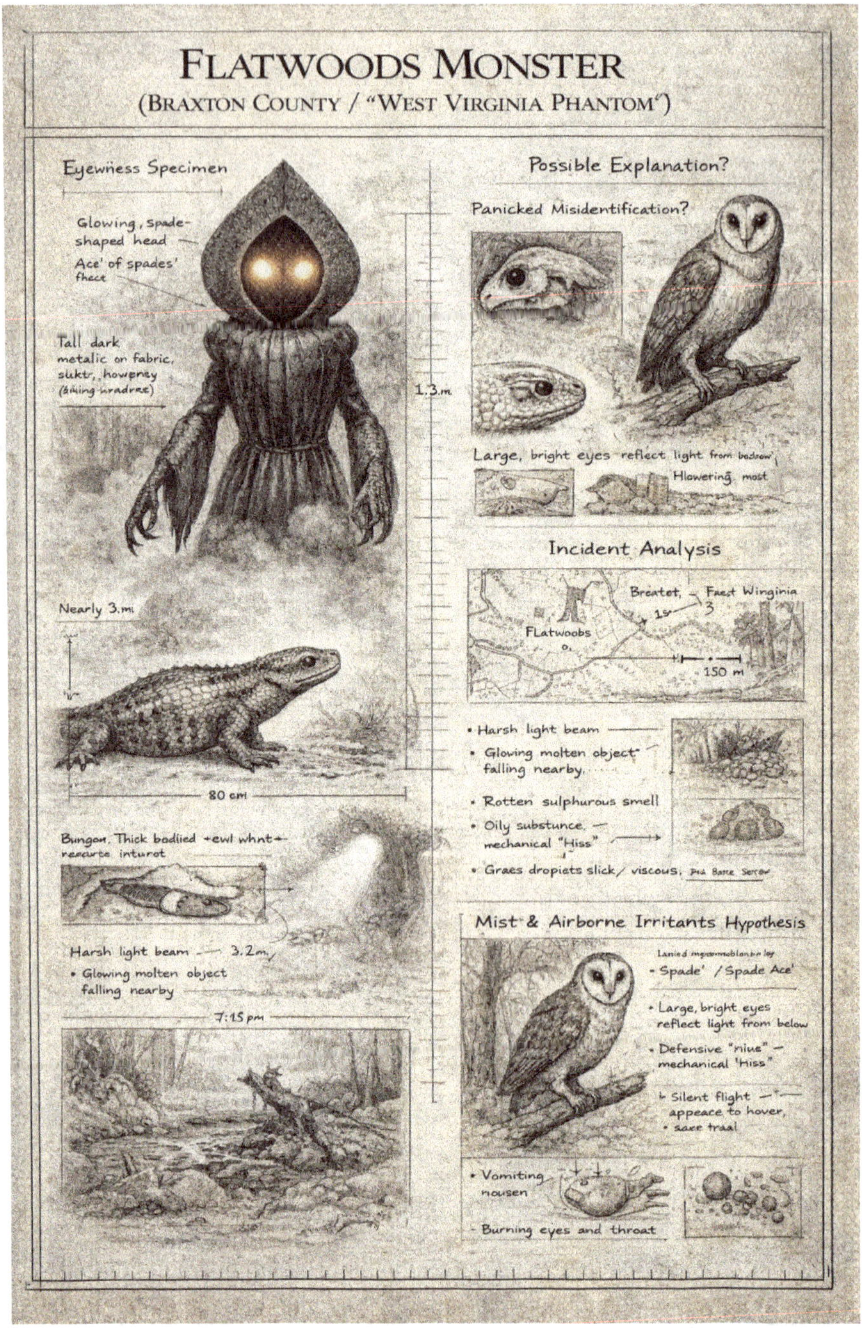

Artificially generated image for illustrative and narrative purposes

Thylacine (Tasmanian Tiger)

In an atlas populated by creatures whose existence rests on faith or speculation, the thylacine occupies a unique—and painful—position: that of the most eagerly awaited "Lazarus." Here we are not in the realm of myth, but in recent historical record. Thylacinus cynocephalus ("dog-headed pouched one") is not a folkloric invention; it was an apex marsupial predator, a masterpiece of convergent evolution that, despite not being related to canids, developed a morphology strikingly similar to that of a wolf in order to occupy the same ecological niche. Until the early twentieth century, the "Tasmanian Tiger" was a real and physical presence in the forests of the Australian island, with its tawny coat and the distinctive dark stripes across its back that earned it its feline nickname.

Its descent into oblivion stands as an indictment of humanity. Unjustly labeled a sheep-killer by European settlers, the thylacine became the target of a systematic eradication campaign fueled by government bounties. The last known specimen—a male named Benjamin—died of cold and neglect at the Hobart Zoo on September 7, 1936, after being locked out of its shelter on a freezing night. With him, the species was considered functionally extinct. And yet, it is precisely after Benjamin's death that the cryptid was born. Since 1936, more than four thousand sightings have been reported in Tasmania alone, and just as many on mainland Australia (where the animal had vanished thousands of years earlier). Rangers, hikers, and tourists continue to describe fleeting encounters with an animal "like a dog, but with a stiff, striped hindquarters," moving with a peculiar, somewhat awkward gait distinct from the fluid motion of a dingo or fox.

Science, typically skeptical, keeps a door slightly ajar in this case. Tasmania contains vast, impenetrable wilderness areas—such as the Tarkine region—where a shy, nocturnal animal could theoretically survive far from human detection. Over recent decades, multiple university-led expeditions have deployed hundreds of camera traps. While none has yet captured the clear, definitive image the world awaits, some have recorded ambiguous shapes and fleeting figures that sustain doubt. The thylacine has thus become a global symbol of "de-extinction": not only are people searching for it alive in the wild, but active efforts are underway to bring it back. Advanced genetic projects, including those conducted by laboratories in Australia and the United States (such as Colossal Biosciences), are attempting to sequence the thylacine genome extracted from museum-preserved specimens, with the aim of "resurrecting" it through genetic editing of related marsupials like the fat-tailed dunnart.

Whether it survives hidden in the bush or returns to life in a laboratory, the thylacine represents the ultimate frontier of cryptozoology. Unlike Bigfoot or Nessie—whose discovery would rewrite zoology—the rediscovery of the thylacine would constitute a moral redemption. Its "hunt" is driven not by fear of a monster, but by collective guilt and the desperate hope that extinction may not, after all, be forever. As noted in the methodological preface regarding the concept of the "Lazarus Taxon," the return of this species would confirm that nature possesses reserves of resilience that scientific arrogance too often underestimates.

1 Newhaven Track, Meunna TAS 7325, Australia	41°04'52"S 145°26'30"E	Last known footage of the Thylacine (1933)
Thylacine Videos and Documentaries	Thasmanian Museum	

Artificially generated image for illustrative and narrative purposes

Artificially generated image for illustrative and narrative purposes

APPENDIX.
The Great Absentees

As we close this atlas, attentive readers may notice a few conspicuous omissions. Where are the figures that dominate contemporary horror and online discussions? Why are Slender Man, werewolves, or the "Greys" of Roswell absent? Their exclusion is not an oversight, but the result of a rigorous methodological criterion that distinguishes cryptozoology—however speculative—from pure mythology, horror fiction, and classical ufology.

To understand the boundaries of this volume, we must return to the definition offered by Bernard Heuvelmans: the study of "hidden animals." The key word is animals. Cryptozoology postulates the existence of biological creatures—made of flesh, bones, and DNA—that live, feed, reproduce, and die according to the ecological laws of Earth.

The first group of exclusions includes monsters of "Digital Folklore," such as Slender Man or Siren Head. These figures are fascinating examples of how fear evolves in the age of the Internet, but we know with absolute certainty that they are works of fiction. Slender Man—the tall, faceless figure in a dark suit—did not emerge from sightings in ancient forests, but from a Photoshop contest on the Something Awful forum in 2009, created by user Eric Knudsen (Victor Surge). It is a case of artificial mythology whose author and date of origin are known. Including it in an investigation of potentially real animals would betray the pact with the reader, conflating the search for biological truth (as in the cases of the thylacine or the coelacanth) with pure literary invention.

The second boundary is drawn against the supernatural. Vampires, werewolves, and zombies belong to the realms of magic, curse, and theology—not zoology. A werewolf that transforms under the full moon violates the laws of mass conservation and evolutionary biology; a Bigfoot, however elusive, is hypothetically a primate that obeys anatomical rules. This book deals with the biologically possible, not the magically impossible. Even creatures like the Chupacabra or the Mothman, though tinged with paranormal overtones, are included because their initial phenomenology—livestock predation, avian characteristics—permits biological or behavioral analysis, even if only to debunk them and trace them back to natural causes such as mange or misidentified nocturnal birds.

Finally, there is the subtle line dividing cryptozoology from ufology. While cases like the Flatwoods Monster or the Puerto Rican Chupacabra brush against alien themes, they present a "monstrous" physicality that interacts with Earth's environment. Classical aliens—the Greys, the Nordics—and their spacecraft, however, belong to the domain of exobiology and extraterrestrial technology. Cryptozoology looks downward, toward the mud, caves, and abyssal depths of our own planet; ufology looks upward. This volume is a tribute to Earth and its last secrets. It is an invitation to consider that, despite satellites and global mapping, our world may still host co-inhabitants who have no intention of being found—hidden not in another dimension, but simply behind the next tree.

The Difference Between Cryptozoology and Pseudoscience

International Society of Cryptozoology

POINT
PLEASANT

Legal Notice

Photo Credits

The images published in this volume are used in compliance with applicable copyright laws and international treaties, including the Berne Convention for the Protection of Literary and Artistic Works, as well as relevant provisions of United States copyright law (Title 17, U.S. Code), including the doctrine of fair use under Section 107.

Most of the images reproduced in this book fall into one or more of the following categories:
• Images sourced from Wikimedia Commons and used in accordance with the terms of their respective Creative Commons licenses (such as CC BY, CC BY-SA, CC0, etc.); where required, author attribution and source information have been provided.
• Photographs and documentary images that are in the public domain due to expiration of copyright protection or lack of copyright eligibility under applicable law.
• Low-resolution reproductions of works that are in the public domain.
• Images reproduced for purposes of research, commentary, criticism, scholarship, illustration, and educational dissemination within a work of study, in accordance with the fair use doctrine (17 U.S.C. §107) and comparable international principles.

Every reasonable effort has been made to identify rights holders and provide proper attribution where required. In the event of any inadvertent errors or omissions, the Author and Publisher are willing to make appropriate corrections or acknowledgments in subsequent editions.

The Author and Publisher assume no responsibility for any improper or unauthorized use of the images contained in this volume by third parties.

QR Codes and Online Content

All QR codes included in this volume, at the time of publication, link to materials and content that are publicly accessible online. However, the Author and Publisher cannot guarantee the continued availability, accuracy, or permanence of external websites or linked content.
The Author and Publisher do not control, and are not responsible for, the ownership status, legality, or subsequent modification of third-party content accessible through external links.
The Author and Publisher disclaim any liability for changes, removals, broken links, or modifications to web addresses occurring after the printing or distribution of this volume.

Visio Series

Visual Explorations
Art, Architecture, Fashion, Cinema, and Visual Culture

Discover all our publications:

Il Disoriente
Great classics of literature

Astra
Essays and stories beyond Earth

Brĕvitĕr
Legal manuals and compendiums

Crimina
Mystery & Noir

First English Books
The best literary classics for English learners

I Grandi della Letteratura
Masterpieces in 6×9 format with high readability

Imaginaria
Science Fiction, Fantasy, and Adventure

Mille Bolle Blu
Children's books

Mysteria
Essays and fiction on mystery themes

Politicamente
Political essays and writings

Retoricamente
Rhetoric, language, and communication

Saggiamente
Essays in humanities and social sciences

Teatro da leggere
Theatrical texts

Università
Academic essays and manuals

Visio
Art, architecture, fashion, cinema, and image culture

www.ingramcontent.com/pod-product-compliance
Lightning Source LLC
LaVergne TN
LVHW022118200926
848481LV00003B/1191
* 9 7 9 1 2 5 9 9 1 8 4 1 3 *